Optimal Investment and Marketing Strategies

SYSTEMS RESEARCH SERIES

Series Editor: Dinesh Verma *(Stevens Inst. of Technology, USA)*

Published:

Vol. 1 System of Systems Collaborative Formation
by Michael J. DiMario

Vol. 2 Optimal Investment and Marketing Strategies
by Ilona Murynets

Systems Research Series — Vol. 2

Optimal Investment and Marketing Strategies

Ilona Murynets

Stevens Institute of Technology, USA

NEW JERSEY • LONDON • SINGAPORE • BEIJING • SHANGHAI • HONG KONG • TAIPEI • CHENNAI

Published by

World Scientific Publishing Co. Pte. Ltd.
5 Toh Tuck Link, Singapore 596224
USA office: 27 Warren Street, Suite 401-402, Hackensack, NJ 07601
UK office: 57 Shelton Street, Covent Garden, London WC2H 9HE

British Library Cataloguing-in-Publication Data
A catalogue record for this book is available from the British Library.

Systems Research Series — Vol. 2
OPTIMAL INVESTMENT AND MARKETING STRATEGIES

ISBN-13 978-981-4383-26-4
ISBN-10 981-4383-26-0

Printed in Singapore.

To my Mother

Contents

Chapter 1

Introduction

1.1 Motivation

Innovative technologies have played and continue to play a central role in an extensive growth of new services in the telecommunication industry. Today almost all US households are subscribed to at least one telecommunication service such as fixed-line or wireless phone, cable or satellite television, DSL, cable or broadband internet, in-store or by-mail DVD rental, etc. Over the past years, post-production maintenance services have also grown considerably. Especially they are important in defense and airline industries supporting equipment with extended life cycles. Thus, investment and marketing strategies for technologically innovative and post-production services is a subject of significant interest in both industry and academia.

Service introduction requires multiple marketing and management decisions to be made well in advance of the service's launch and throughout the service's entire life-cycle. Obviously, these decisions need models for accurate prediction of service demand. Often area experts underestimate innovative technologies and services, which later prove to be successful. For example, in 1878 William Preece, a chief engineer in the British Post Office, said that there was no need for telephony, since there were "plenty of messenger boys" [Preece (1878)]. A memo at Western Union 1878 stated that "the telephone could not be seriously considered as a means of communication and it did not have any value" [Olshansky and Dossey (2003)]. In 1946 Darryl Zanuck, a famous 20th Century Fox movie producer, said that "television won't be able to hold on to any market it captures after the first six months" and that "people will soon get tired of staring at a plywood box every night" [Schubin (2008)]. A similar forecast was given in 1948 by Mary Somerville, a pioneer of radio educational broadcasts, who said that "television won't last. It's a flash in the pan" [Somerville (1948)].

Nowadays, it is meaningless to argue that telephony and television were adopted successfully by customers. Telecommunication services play an increasingly important role all around the world. They have become an integral part of daily life of an average person. Recent studies suggest that an average American watches TV for more than 4 hours each day [California State University Northridge online (2009)] and uses phone for more than 13 hours per month [Leo (2006)].

In the telecommunication industry, new technologies and services constantly substitute the legacy ones. Well-known examples of such substitutes include: (i) wireless internet as a substitute for a wired internet (ii) Voice-over-Internet Protocol (VoIP) as a substitute for fixed-line telephony, (iii) content distribution via downloading as a substitute for distribution via physical storage such as CDs, DVDs and books, to name just a few. New substitutes inevitability affect supply and demand curve of corresponding legacy products and services. For example, Sound Partners Research Group [Sound Partners Research (2008)] predicted that by 2015, cellular VoIP will carry 28% of all fixed and mobile voice minutes in the USA and 23% in Western Europe. According to TeleGeography's May 2008 press release [TeleGeography, a research division of PriMetrica, Inc. (2008)], since the beginning of 2005, the three Regional Bell Operating Companies, AT&T, Verizon and Qwest, have lost 17.3 million residential telephone lines equivalent to billions of dollars in revenue. In contrast, VoIP service providers have gained 14.4 million new customers.

Technological substitutes significantly complicate investment and marketing strategies. If a provider of legacy services introduces a technologically innovative service, its strategy should neither undermine the market value of the new service nor cannibalize profits of the legacy service. In general, service substitution can be analyzed on a macro level (company or country) as well as on a micro level (individual customer). Successful decision making should consider both levels. Macro level analysis is used for long-term strategic planning, and most of the existing research in this direction is based on Bass's model [Bass (1969)]. However, companies cannot rely only on the macro level analysis. Usually, the success of companies' short-term decisions depends on a detailed insight of the behavior of their existing customers, achieved only by micro level analysis [Kumar and Petersen (2008)]. Rogers [Rogers (1976)] was one of the first researchers to discuss the substitution of a legacy product by a new one on the individual customer level.

Despite the growing role of the service sector, existing literature does not adequately address the diffusion of new subscription services and optimal pricing strategies for service providers [Mesak and Darrat (2002)]. Majority of the diffusion models analyzes technological innovations in connection with durable goods rather than subscription services [Mahajan and Muller (1979)]. There are several similarities and differences in diffusion processes of durable goods and subscription services. Adopters of durable goods as well as adopters of subscription services are largely influenced by advertisements and word-of-mouth effect. Also, adopters of any innovation are "psychographically described as price-conscious shoppers" [Warren *et al.* (1989)], and consequently, the rate of either durable good or service adoption strongly depends on its price and/or subscription fee. The first essential difference is that models describing diffusion of services should simultaneously account for both an activation fee and a subscription fee. The second important difference is that at each time period, the revenue of a subscription service is generated not only by new adopters but also by adopters from the previous periods [Nagle (1987)], and the third essential difference is that customer retention and discontinuance (churn), being key factors in diffusion of subscription services [Kim *et al.* (2009)], play no role in diffusion of durable goods. These differences explain why existing diffusion models and pricing strategies for durable goods are not applicable to subscription services.

To decrease the churn rate and to enhance the word-of-mouth effect, service providers should maintain high service quality and should improve customer support. It is well-known that quality (reliability) of a new service is highly correlated with initial investment. However, an exact relationship between the investment and service quality in each particular case is not obvious. Recently, supplier-customer relationships for a post-production maintenance service have shifted from traditional maintenance service contracting (TMSC) to performance-based contracting (PBC). Under PBC, a customer pays a fixed fee upfront and a supplier is responsible for all maintenance service costs. PBC encourages suppliers to invest in system reliability, which reduces maintenance costs and increases operational availability. On the other hand, high investment increases service fees, and thus, each company maximizes its profit by trading off an investment, system reliability and service fees. Most of research publications on PBC are qualitative [Keating and Huff (2005); Kim *et al.* (2007); Sols *et al.* (2007, 2008)] and none of the existing works develop optimal investment and pricing strategies for PBC.

In summary, many aspects of investment and marketing strategies for technologically innovative services have been largely unaddressed. Specifically, there is a lack of rigorous analysis and models for

(i) customers' migration from legacy services to new substitutes;
(ii) identification of customers for targeted marketing campaigns;
(iii) diffusion of new services on monopolistic and duopolistic markets;
(iv) optimal service investment and pricing strategies.

1.2 Contribution

The book addresses problems (i)-(iv) above, specifically:

(1) The book presents a ***model for customer identification for targeted marketing campaigns*** using data of customers' migration from a legacy service to a new technological substitute on the individual customer level. The model analyzes a consumption share of the new service and forecasts migration behavior of each customer. It also segments customers based on their migration patterns (migrated customers, customers which are going to migrate, retained customers, etc.) and projects new service's demand for the future. A case study illustrates the model based on the real-life company data. The book discusses the main factors influencing marketing strategies for new telecommunication services in different demographical groups of customers and provides managerial suggestions.
(2) The book develops ***analytical single-period models for optimal pricing strategies for new services on monopolistic and duopolistic markets***. The model for a monopolistic market assumes Weibull distribution of customers' reservation prices and maximizes service provider's profit with respect to a service fee. The model for a duopolistic market assumes bivariate gamma distribution of customers' reservation prices for two competitive services and identifies a Nash equilibrium for optimal subscription fees of both competitors.
(3) The book develops an ***analytical model for dynamic optimal pricing strategies for new services on a monopolistic market***. The model accounts for advertising and word-of-mouth effects, customers' churn and customers' willingness to pay for the service. It finds a closed-form solution for an optimal dynamic pricing strategy that includes a one-time activation fee and a subscription fee maximizing the total profit of the service provider over a given planning horizon. The optimal activation fee is low at the introductory stage and gradually increases over the planning horizon, whereas the optimal subscription has an opposite behavior. The optimal pricing strategy is illustrated numerically. The demand for the service is forecasted for each time

moment over the entire planning horizon based on the optimal pricing strategy.

(4) The book develops an ***analytical model for optimal dynamic pricing strategies of two competitive services on a duopolistic market.*** The model assumes, that based on service fees and customers' willingness to pay for each service, a customer either subscribes to a single most preferable service, or does not subscribe to any of them. It also accounts for churn of unsatisfied customers, advertising and word-of-mouth effects. The model shows that there is a Nash equilibrium for optimal subscription fee strategies of two competitors. Those strategies are analyzed numerically. The demand for each service is forecasted for each time moment over the entire planning horizon based on the optimal pricing strategies.

(5) The book develops a ***decision-theoretic model for optimal investment and pricing strategies for a new performance-based post-production maintenance service contract*** that maximize profit of a supplier. The model analyzes reliability as a function of investment, variability of a cost per failure and customers' willingness to pay for the contract depending on contract's length. Numerical examples illustrate how optimal strategies and optimal profit depend on the contract's length, potential market size, expected cost per failure and on other parameters of the model.

1.3 Organization of the Book

Organization of the book is shown on Figure 1.1. Chapter 2 discusses the existing research on new service diffusion and optimal investment and pricing strategies for technologically innovative subscription services and post-production services. Chapter 3 presents the model for identification of customers for targeted marketing campaigns based on analysis of customers' migration from a legacy service to a new technological substitute on the individual customer level. Chapter 4 considers the single-period optimal service pricing models for monopolistic and duopolistic markets. Chapter 5 develops the analytical model for dynamic pricing for a technologically innovative service on a monopolistic market. Chapter 6 develops the analytical model for dynamic pricing of two competitive technologically innovative services on a duopolistic market. Chapter 7 develops optimal investment and pricing strategies for performance-based post-production services. Chapter 8 concludes the book.

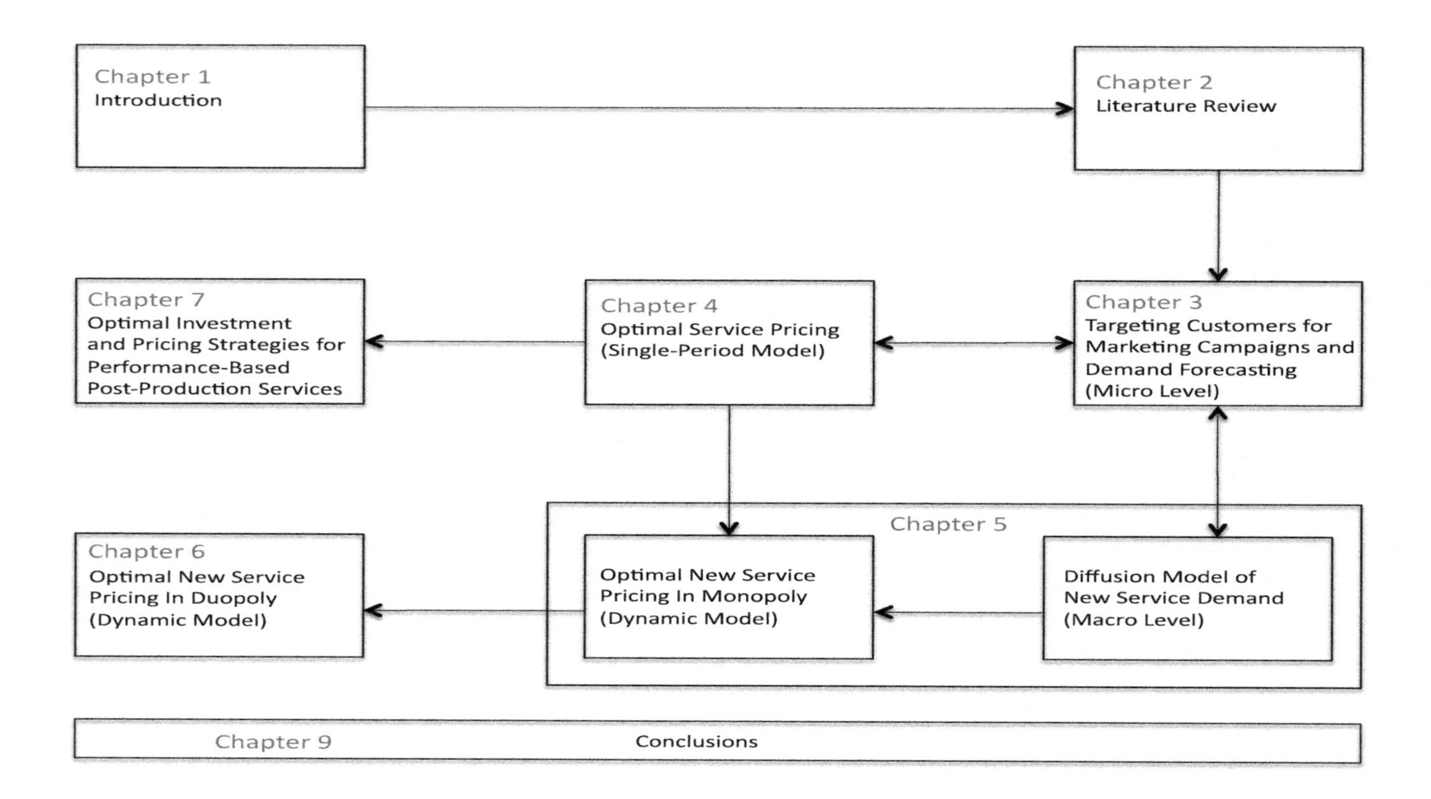

Fig. 1.1 Organization of the book

Chapter 2

Literature Review

2.1 New Service Diffusion

Adoption of innovative products has been studied qualitatively and quantitatively on both macro and micro levels. Rogers [Rogers (1962)] was arguably the first who studied the substitution of a legacy product by a technologically innovative product on the individual customer (micro) level. As a plausible approach, he suggested surveying customers at random and studying their personal and demographical characteristics along with their migration behavior [Rogers (1976)]. Rogers classified adopters of an innovation into five categories based on the fact that some customers are more likely to adopt a new product earlier than others. These categories are innovators, early adopters, early majority, late majority and laggards, see Figure 2.1. Innovators constitute a category of individuals, who are willing to take risks and to try a new product right after its emergence on a market. This category includes young, socially active, educated and financially successful individuals. Early adopters are a well-informed category of customers that adopt a new product based on successful adoption experience of innovators. These individuals are opinion leaders and have very similar demographical characteristics to innovators category. Individuals in the early majority category adopt a new product under influence of early adopters and have an above-the-average social status. Individuals in the late majority category adopt an innovation after the average member of the society and they are characterized by skepticism and a below-the-average social status. Individuals in the laggards category are the last to adopt an innovation if ever, have the lowest social status and are the oldest among all adopters. However, it is not necessary that innovators or early adopters of one product will become innovators or early adopters for the

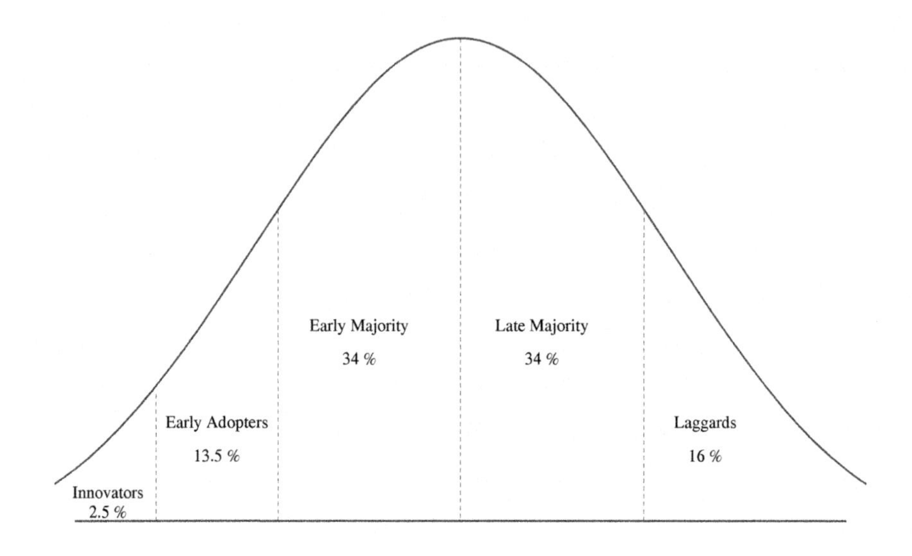

Fig. 2.1 Rogers' adopter categories (adopted from [Mahajan *et al.* (1995)])

others. Constantiou and Kautz [Constantiou and Kautz (2008)] surveyed customers in Denmark and explored affects of customers' perceptions of key economic factors on the substitution of traditional telephony by IP telephony. Similarly, IBM surveyed 600 consumers in the United States, China and the United Kingdom and studied their mobile Internet preferences [Journal (2008)] and Allenet and Barry [Allenet and Barry (2003)] surveyed over 1000 French pharmacists and studied substitution of brand drugs by generic drugs. However customer's surveying is extremely time consuming, expensive and customers sometimes provide hasty and misleading responses [Marsland *et al.* (2000)].

Johnson and Bhatia [Johnson and Bhatia (1997)] used regression analysis to study substitution on the land mobile radio (wireless) market. They suggested five stages of substitution on the micro level: awareness, interest, evaluation, trial and adoption or rejection. Meuter *et al.* [Meuter *et al.* (2005)] introduced multiple and logistic regression models for studying the substitution of clerk-based services by self-service technologies, such as telephone banking, automated hotel checkout, and online investment trading. Prins [Prins (2008)] used the split-hazard approach to model a substitution of fixed-lined phones by a new mobile service in the Dutch consumer market. A partitioned fuzzy integral multinomial logit model was applied to studying substitution on the Taiwan's Internet telephony market [Tseng and Yu (2004)]. The existing literature suggests that risk, complexity and ease-of-use of a new product along with the demographic characteristics of customers (for example, young males with high income) have the highest impact on a success of substitution [Manning *et al.* (1995); Steenkamp and Burgess (2002); Chopra (2005); Nader and Jimenez (2005)]. Table 2.1 presents the summary of the literature analyzing substitution of legacy products by new ones on a micro level.

Quantitative research on diffusion of innovation on a macro level was initiated by a pioneer work of Bass [Bass (1969)] and was aimed to describe the cumulative number of adopters of new durable goods over time. Bass assumed that the rate of sales of a consumer durable good depends on a number of current adopters, the number of potential adopters as well as on an external influence (advertisement) and on an internal influence (word-of-mouth). The model is given by the differential equation

$$\dot{x}(t) = (M - x(t))(a + bx(t)), \tag{2.1}$$

where $x(t)$ represents the cumulative sales to time t, $\dot{x}(t)$ is a derivative of $x(t)$, which represents a change in the cumulative number of adopters at time t, M is a total number of potential adopters, a is a coefficient of innovation and b is a coefficient of imitation. Bass classified all adopters into two categories: innovators and imitators. Individuals in the innovators category adopt a new product under influence of advertisement while individuals in the imitators category adopt a new product as a result of a word-of-mouth effect.

Mahajan *et al.* [Mahajan *et al.* (1990)] reviewed more than a hundred different extensions of this model. For example, one of the extensions analyzed several generations of substitutions on Korea's mobile communication service market [Kim *et al.* (2000)]. Also using an S-shaped curve, Fisher and Pry [Fisher and Pry (1971)], suggested that if a new technology reaches 10 percent market share it will conquer the whole market. However, the S-shaped curve, introduced by Bass, is an increasing function that may capture only successful substitutions, when new products are accepted by customers and it is inapplicable for analysis of unsuccessful substitutions. Several Bayesian statistical models were also used to forecast the substitution of old durable products [Lenk and Rao (1990)] and old movies [Neelamegham and Chintagunta (1999)] by new ones. Some studies suggested that the rates of substitution of a legacy product by new technological substitutes depend on cultural values of each nation [Dekimpe *et al.*

(1998); Tellis *et al.* (2003); Steenkamp *et al.* (1999)]. The summary of the literature on the macro level is presented in Table 2.2.

The methods used for the modeling of a diffusion of a new product include descriptive analysis, linear, multiple and logistic regression, survey analysis, split-hazard approach and Bayesian approach, etc. However, the existing literature lacks methods that are based on the analysis of historical usage to forecast the migration of individual customers from legacy services to new ones caused by the emergence of new technologies. There are no tools available to companies to forecast how many customers are likely to migrate from a legacy product to a a new one and to segment their customers based on customers' migration behavior in order to target them with individual marketing campaigns.

2.2 Service Pricing in Monopoly

Aforementioned research did not consider any decision variables. Robinson and Lakhani [Robinson and Lakhani (1975)] were the first to incorporate a price as a decision variable into the Bass model (2.1) and thus initiated a new stream of research focused on development of dynamic optimal pricing strategies for maximization of the total profit over the planning horizon by a monopolist company. They assumed that the rate of a durable good adoption is a decreasing function of the good's price. Robinson and Lakhani's [Robinson and Lakhani (1975)] model is given by the optimal control problem

$$\begin{aligned} &\max_{p(t)} \int_0^T (p(t) - c(t))\dot{x}(t)dt \\ &\text{s.t.} \quad \dot{x}(t) = (M - x(t))(a + bx(t))\,\mathrm{e}^{-p(t)}, \quad x(0) = x_0, \end{aligned} \tag{2.2}$$

where $p(t)$ is a price, $c(t)$ is cost, x_0 is initial level of adoption and $[0, T]$ is a planning horizon. Papers addressing optimal dynamic pricing of new consumer goods by monopolists include Dolan and Jeuland [Dolan and Jeuland (1981)] and Jeuland and Dolan [Jeuland and Dolan (1982)], Horsky and Simon [Horsky and Simon (1983)], Teng and Thompson [Teng and Thompson (1983)], Dockner and Jorgensen [Dockner and Jorgensen (1988)], Feng and Gallego [Feng and Gallego (2000)], Ruiz-Conde, Leeflang and Wieringa [Ruiz-Conde *et al.* (2006)].

Most of the diffusion models of innovation are focused on durable goods and do not take into account re-purchasing and customer attrition. Dodds

Table 2.1 Summary of the related literature on the micro level

Literature	Research Approach	Application
Johnson and Bhatia (1997)	Regression analysis	Land mobile radio (wireless) communication market
Neelamengham and Chintagunta (1999)	Bayesian model	Movie industry
Allenet and Barry (2003)	Descriptive analysis	Generic Drugs (French pharmacists market)
Teng and Yu (2004)	Fuzzy integral multinomial logit model	Internet Telephony (Taiwan market)
Meuter, Bitner, Ostrum, Brown (2005)	Multiple regression and logisti regression	Self-serving technologies
IBM (2008)	Survey analysis	Mobile Internet (US,China, UK markets)
Prins (2008)	Split-hazard approach	Mobile telephony (Dutch market)
Constantinou and Kautz (2008)	Survey analysis	IP telephony

Table 2.2 Summary of the related literature on the macro level

Literature	Approach	Application
Bass (1969)	Bass diffusion model	Consumer durables
Fisher and Pry (1971)	S-curve model	Forecasting of technological opportunities
Lenk and Rao (1990)	nonlinear regression, Bass model	Consumer durables
Dekimpe, Parker and Sarvary (1998)	Bass Diffusion model	Cellular telecommunication market
Kim , Seo and Lee (1999)	Bass model extension	Mobil telecommunication market in Korea
Steenkamp. Hofstede, Wedel (1999)	Regression	Technological innovativeness
Tellis, Stremerch, Yin (2003)	Parametric hazard model	Consumer durables
Nader and Jimenez (2005)	Descriptive analysis	Electronic mail alternatives
Chopra (2005)	Descriptive analysis	Electronic mail alternatives

[Dodds (1973)] was arguably the first who applied the Bass model (2.1) to a subscription service. The model provided a reasonably good long-term forecast for the cumulative number of cable television service adopters based on early data. Kim, Mahajan and Srivastava [Kim *et al.* (1995)] used a technological substitution model of Fisher and Pry [Fisher and Pry (1971)], corresponding to the Bass model (2.1) with $M = 1$ and $a = 0$, to forecast diffusion of a cellular communication service. However, (2.1) does not include product's price and accounts for neither subscription fee nor customer attrition critical for developing optimal pricing strategies. The diffusion model of Libai, Muller, and Peres [Libai *et al.* (2009)] for a subscription service addresses this deficiency only partially: it accounts for the effect of customer attrition but incorporates no price variable.

An optimal pricing policy of Fruchter and Rao [Fruchter and Rao (2001)] includes an ongoing membership fee (e.g. monthly payment) and a usage price (price per minute) that maximize the sum of discounted profits. The membership fee is set low at the beginning and then increases with network size (penetration strategy), while the usage price is set high and then decreases over time (skimming strategy). Mesak and Darrat [Mesak and Darrat (2002)] extended the Bass model by introducing a price variable and a relationship between the adoption process of consumers and the adoption process of retailers. According to their pricing strategy, the subscription fee for new subscription services decreases during an introductory period and increases thereafter.

In summary, a diffusion of new services and optimal new service pricing have not been studied extensively in the literature. Some of the works accounted for the effect of a price on a service diffusion and developed optimal pricing policies for service providers, while did not pay any attention to a customer attrition. Others, accounted for the attrition effect, while did not consider a price variable at all. Also, there is no work accounting for both a subscription fee and an activation fee charged by a service provider. A detailed analysis of the existent literature on diffusion of services and service pricing is provided in Tables 2.3(a) and (b).

2.3 Service Pricing in Duopoly

For the past three decades, pricing of substitutable products under competition has been extensively studied. Kreps and Scheinkman [Kreps and Scheinkman (1983)] determined optimal prices and capacities for two

Table 2.3(a) Related literature on diffusion of services and service pricing (Part 1)

Literature	Diffusion Model	Profit Maximization	Objective Function	Decision Variable	Churn	Optimal Strategy	Application
W. Dodds "Application of the Bass Model in Long-Term New Product Forecasting" **(1973)**	$\frac{dx(t)}{dt} = (M - x(t))(a + bx(t))$	No	Not applicable	None	No	Not applicable	Cable TV service
A.Dhebar, S. Oren Optimal Dynamic Pricing for Expanding Networks " **(1985)**	$\frac{dx(t)}{dt} = F(M(p(t), x(t), x(t))$	Yes	$\int_0^\infty e^{-rt}(p(t)x(t) - c(x(t)))dt$	Price	No	Involve a "jump" in the subscription price from zero to the optimal steady-state value	Not included
N. Kim, V. Mahajan, R. Srivastava "Determining the Going Market Value..." **(1995)**	$\frac{dx(t)}{dt} = (1 - x(t))bx(t)$ $x(t) = 0$	No	Not applicable	None	No	Not applicable	Cellular Telecom Industry.

Table 2.3(b) Related literature on diffusion of services and service pricing

Literature	Diffusion Model	Profit Maximization	Objective Function	Decision Variable	Churn	Optimal Strategy	Application
G.Fruchter, R. Rao "Optimal Membership Fee and Usage Price over Time for a Network Service" **(2001)**	$\frac{dx(t)}{dt}=(M-x(t))(a+bx(t))k^{-\alpha_1}-\delta x(t)(p+c_1)^{-\alpha_2}$ $D(x(t),q(t))=m_0(d+x(t))^{\alpha_3}k^{-\alpha_4}$	Yes	$\int_0^{\infty} e^{-rt}x(t)((p(t)-c_1)+(k(t)-c_2)D(x(t),p(t)))dt$	Price	Yes	Optimal policy consists of a penetration strategy for the membership fee and the skimming strategy for the usage fee,	Not included
H. Mesak, A.Darrat "Optimal pricing of new subscriber services under ..." **(2002)**	$\frac{dR(t)}{dt}=(N(t)-R(t))(a_2+b_2R(t))$ $\frac{dx(t)}{dt}=(M(t)-x(t))(a_1+b_1x(t))h(p(t))$ $N(t)=\alpha_1+\beta_1 x(t)$ $M(t)=\beta_2 R(t)$	Yes	$\int_0^{T} e^{-rt}((p(t)-c(t))x(t)-S(R(t)))dt$	Price	No	Increasing the subscription fee for new subscriber services over time, following an introductory period in which the fee may be decreasing	Not included
B. Libai, E.Muller, R.Peres "The Diffusion of services" **(2009)**	$\frac{dx(t)}{dt}=(M-x(t))\left(a+\frac{b(1-\delta)x(t)}{M}\right)-\delta(x(t)$	No	Not applicable	None	Yes	Not applicable	Cell phone

perfectly substitutable products, assuming that the demand for each product depends on its price. Staiger and Wolak [Staiger and Wolak (1992)] developed optimal pricing and capacity strategies of two firms producing the same product and facing stochastic demand. They assumed that the firms choose capacities of their products first and then simultaneously set their prices. Birge *et al.* [Birge *et al.* (1998)] examined capacity and price strategies for a company producing two substitutable products with independent uniformly distributed demands in two cases: when the company maximizes its overall profit and when it maximizes profits of the products separately. Using a game-theoretical approach, Jing Zhao *et al.* [Zhao *et al.* (2008)] explored pricing strategies of two competitive retailers providing the same two types of substitutable products. In all these works, considered models are static and do not account for diffusion of new products or services.

A stream of research on dynamic pricing strategies in duopolies was initiated by Dockner [H. Steckhalm et. al. (1984); Dockner (1985)] who generalized Robinson and Lakhani's [Robinson and Lakhani (1975)] model. He assumed that a diffusion of a new product was driven only by an innovation effect and did not incorporate a word-of-mouth effect. The model resulted in optimal prices that decrease with time. Fleichtinger and Dockner [Feichtinger and Dockner (1985)] found a Nash equilibrium for optimal product pricing strategies of multiple profit-maximizing firms. They assumed that the number of customers is a decreasing function of product price. The major drawback of Fleichtinger and Dockner's [Feichtinger and Dockner (1985)] model is that it does not consider an inflow of customers. Chintagunta and Rao [Chintagunta and Rao (1996)] used a game-theoretic approach to develop dynamic pricing strategies of two firms on a duopolistic market. They assumed that preferences for brands evolve over time and that demand for a brand is given by a logit model. Granot *et al.* [Granot *et al.* (2007)] analyzed a duopoly competition between providers of an identical product. They assumed that in each period, a customer visits only one of the retailers. If the price charged by the retailer is below customer's reservation price, the customer purchases the product. Otherwise, in the next period, the consumer visits the other retailer.

In summary, the existing literature on dynamic pricing in duopolies focuses mostly on goods. Models for optimal dynamic service pricing in duopoly should assume that customers adopt a service with the highest utility and discontinue a service if its fee is higher than competitor's.

2.4 Optimal Investment and Pricing Strategies for Performance-Based Post-Production Service Contracts

This section presents a review of relevant literature on optimal investment and pricing strategies for performance-based and traditional post-production service contracts and on reliability design. Performance-based contracting (PBC) has recently emerged as an important subject of discussion in defense and commercial sectors, and academic research in this direction is in its infancy. The existing literature provides mostly qualitative insight into current practices and implications of PBC including multiple government-issued guidebooks for suppliers [Defense Acquisition University (2005a,b)]. FCS Group consulting [for Office of Financial Management (2005)] conducted a literature review and surveyed several agencies and local jurisdictions that have implemented PBC on the best practices and trends in PBC. They found out that suppliers had a number of management issues and difficulties related to implementing PBC. Keating and Huff [Keating and Huff (2005)] suggested that in the aerospace industry PBC shifted risk from a customer to a supplier. On the quantitative side, Sols *et al.* [Sols *et al.* (2008)] developed an n-dimensional effectiveness metric-compensating reward scheme in PBC; Nowicki *et al.* [Nowicki *et al.* (2008)] examined inventory allocation under PBC; and Kim *et al.* [Kim *et al.* (2007)] developed a principle-agent model to study the implications of PBC, by analyzing allocation of performance requirements and risk sharing when a single customer is contracting with a collection of suppliers. Unfortunately, the existing literature on PBC offers only little guidance for contract execution. The following questions remain unaddressed. What is an optimal contracting period? How to price a contract? Is it profitable to invest in reliability design? How much to invest?

Traditional post-production service contracting (TPSC) has been extensively studied in the literature [Sherif and Smith (1987); Stremersch *et al.* (2001); Levery (2002)], however the existing models for TPSC are inapplicable for performance-based contracting, since they do not optimize pricing and investment strategies simultaneously. For example, Murthy and Yeung [Murthy and Yeung (1995)] presented game-theoretic optimal maintenance strategies for a customer and an independent service provider. They assumed that the customer determines frequency of maintenance services and the provider specifies costs and a schedule for ordering spare parts. Murthy and Asgharizadeh [Murthy and Asgharizadeh (1999)] and Asgharizadeh and

Murthy [E. Asgharizadeh (2000)] developed game-theoretic models under the assumption that customers have two options: accepting a fixed-priced contract (with contract price and repair costs determined by the provider) or repairing equipment at own expense. Jackson and Pascual [Jackson and Pascual (2008)] determined the optimal price and number of clients for maintenance service contracts that maximize the profit of a service provider. Kim *et al.* [Kim *et al.* (2009)] evaluated provider's trade-off between investing in reliability improvements and stocking spares.

System reliability [Murthy and Blischke (2006)] is central to any performance-based post-production contractual arrangement. It is defined as the probability that the system will perform its intended function for a specified time period when operating under normal (or stated) environmental conditions. In literature, the notions of reliability and quality are often used interchangeably. Majority of research on reliability (quality) design optimizes trade-off between system reliability (quality) and market entry timing [Lilien and Yoon (1990)]. For example, Deshmukh and Chikte [Deshmukh and Chikte (1977)] presented a semi-Markov decision model for optimal funding of a product quality improvement project and determined the optimal time to terminate the project. They assumed that a product's profit depends on the product's quality and quality of all competing products. Cohen *et al.* [Cohen *et al.* (1996)] developed a multistage model for a product quality improvement process optimizing a time-to-market and a performance target. Levesque [Levesque (2000)] explored effects of funding on product quality and developed an analytical framework for optimal stopping rule for the development of the new product. Murthy *et al.* [Murthy *et al.* (2009)] developed a qualitative framework allowing manufacturers to achieve an optimal trade-off between an investment and a cost of consequences of inadequate product reliability. To the best of our knowledge, there is no research work developing a model for reliability design in the context of performance-based contracting.

In summary, there is a lack of quantitative models for optimal investment and pricing strategies related to performance-based post-production service contracts, and models for optimal pricing of traditional maintenance services do not optimize an investment in reliability design.

Chapter 3

Migration of Customers and Targeting Customers for Marketing Campaigns

3.1 Overview

Telecommunication, computer, pharmaceutical and financial industries suffer significant losses from their legacy products and services due to the constant emergence of new technological substitutes. Well-known examples of such substitutes include: (i) content distribution via downloading as a substitute for distribution via physical storage (e.g. CDs, DVDs and books), (ii) Voice-over-Internet Protocol (VoIP) as a substitute for fixed-line telephony and (iii) generic drugs as substitutes for brand drugs, to name just a few. In fact, new substitutes inevitability affect the supply and demand curve of corresponding legacy products and services. For example, Sound Partners Research Group [Sound Partners Research (2008)] predicted that by 2015, cellular VoIP will carry 28% of all fixed and mobile voice minutes in the USA and 23% in Western Europe. Also, according to TeleGeography's May 2008 press release [TeleGeography, a research division of PriMetrica, Inc. (2008)], since the beginning of 2005, the three Regional Bell Operating Companies, AT&T, Verizon and Qwest, have lost 17.3 million residential telephone lines equivalent to billions of dollars in revenue. In contrast, VoIP service providers have gained 14.4 million new customers. In the entertainment service industry, the well-known DVD rental providers Blockbuster and Hollywood Video have lost their customers to Netflix, Apple and multiple other companies providing movies via mail and/or real-time download. As a result, Blockbuster lost $1.2 billion in 2004 and over $500 million in the subsequent year [Reisinger (2008)], and Movie Gallery Inc., the parent of Hollywood Video, filed for bankruptcy in October 2007 [Movie Gallery Inc. (2008)]. Standard & Poor estimated that brand drugs such as Risperdal,

Topamax, Fosamax, Depakote and Zyrtec will loose approximately $20 billions in sales due to migration of patients to generic drugs [Edwards (2008)].

Although, it is common that companies providing legacy products view emergence of new technological substitutes as potential losses, they can also gain from additional opportunities created by these new technologies. To remain competitive in such rapidly changing markets, companies can either improve their legacy products and services and thus, retain the existing customers via marketing campaigns or can upgrade the infrastructure to satisfy customers' demand for new technologies. For example, after Blockbuster realized Netflix was a serious competitor and threat, it improved the existing service by increasing merchandize assortment and in-stock availability, reducing prices, modifying in-store merchandize layout and canceling late fees. However these actions were not sufficient for regaining market share from Netflix [Reisinger (2007)]. Blockbuster realized that embracing the new online technology was necessary to remain competitive. Thus, Blockbuster introduced the "total access" service enabling customers to rent and return DVDs across different channels [Randolph (2007)]. Also, it purchased the digital movie-download service Movielink for real-time downloading of high-definition quality movies and opened self-service DVDs kiosks. These new initiatives and diversification policies have made Blockbuster the first serious rival to the original online market leader, Netflix [Stories (2008)].

Substitution of legacy products by new technological substitutes has been and continues to be an important subject in marketing research, particularly, because decision making relies on its analysis. However, the existing literature lacks analysis of activity profiles in application to customer's migration due to product substitution, including analysis of common migration patterns, causes of migration, identifying segments for marketing campaigns, etc. This chapter bridges the gap. It analyzes customers' migration from a legacy service or product to a new substitute on the micro level, based on new substitute's usage share data. The analysis includes forecasting of future migration behavior of each customer, constructing a confidence interval for the forecast via bootstrapping, estimation of a probability distribution of the new substitute's usage share of the entire customer sample and segmentation of customers based on their migration behavior. The chapter is organized into four sections. Section 3.2 presents an analysis of the customer's migration from a legacy product to a new technological substitute on a customer level (micro level), Section 3.3 considers a case study, Section 3.4 presents practical implications and concludes the chapter.

3.2 Model

This section analyzes customers' migration from a legacy product to a new technological substitute on a micro level. Consider a sample consisting of m random customers, who use at least one of the products: the legacy product or the new technological substitute. Let $\boldsymbol{z}_i = (z_{i1}, \ldots, z_{iT})$ be the historical activity data of customer i, where

$$z_{it} = \frac{y_{it}}{x_{it} + y_{it}}$$

is the customer's usage share of the new substitute at time t with x_{it} and y_{it} representing the customer's usages of the legacy product and its substitute, respectively at time t, for $t = 1, \ldots, T$ and $i = 1, \ldots, m$.

The analysis consists of the following three steps. The first step fits $\boldsymbol{z}_i$ into an individual migration profile f_i. The second step uses f_i to forecast the future migration behavior of customer i, $\widehat{z}_i$ and estimates a probability distribution for $\widehat{z}_i$, $i = 1, \ldots, m$. It also uses the bootstrapping algorithm to construct an α-confidence interval for $\widehat{z}_i$. Finally, the third step segments customers by similar migration profiles for targeted marketing campaigns. The remaining of the section discusses each step in detail.

Profiling

The first step fits the activity data $\boldsymbol{z}_i$ into a migration profile f_i, with the Adbudg S-shaped function

$$f_i(t) = f(a_i, b_i, c_i, d_i, t) = b_i + (a_i - b_i)\frac{t^{c_i}}{d_i + t^{c_i}}, \qquad t = 1, \ldots, T, \quad (3.1)$$

where parameters $0 \leq a_i, b_i \leq 1$, $c_i \in \mathbb{R}$, $d_i \in \mathbb{R}$ minimize the total quadratic error

$$\min_{a_i, b_i, c_i, d_i} \sum_{t=1}^{T} (f(a_i, b_i, c_i, d_i, t) - z_{it})^2, \qquad i = 1, \ldots, m. \quad (3.2)$$

The Adbudg function, introduced by Little [Little (1970)] for advertising budgeting, is flexible enough to describe any increasing or decreasing S-shaped function: $f_i(t)$ is an increasing S-shaped curve on $[0, \infty]$ if $(a_i - b_i)c_i d_i > 0$ and $d_i(c_i - 1)/(c_i + 1) \geq 0$ (see Figure 3.1 (a)) and $f_i(t)$ is a decreasing S-shaped curve on $[0, \infty]$, if $(a_i - b_i)c_i d_i < 0$ and $d_i(c_i - 1)/(c_i + 1) \geq 0$ (see Figure 3.1 (b)). In the first case, $f_i(t)$ has the upper asymptote $f(t) \to a_i$ as $t \to \infty$ and in the second case $f(t)$ has a lower asymptote at $f(t) \to b_i$ as $t \to \infty$. The inflexion point of $f_i(t)$ is at

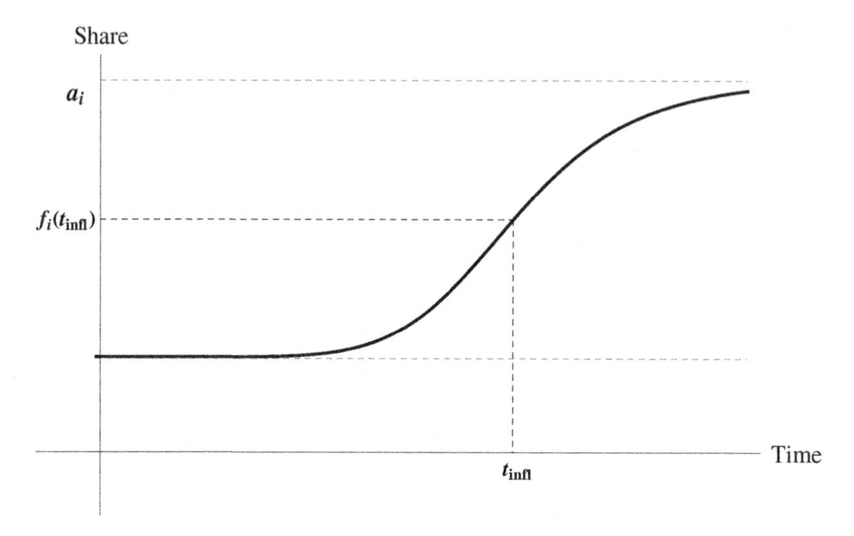

(a) Successful substitution

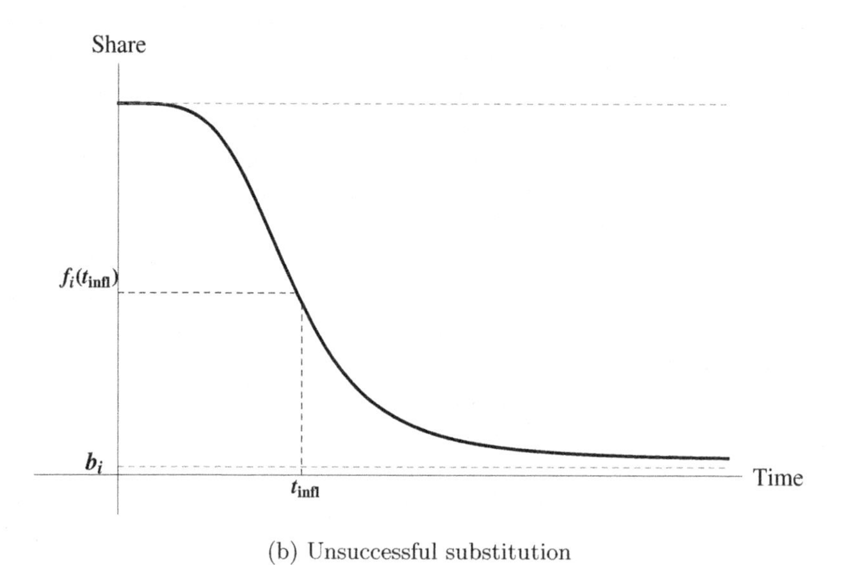

(b) Unsuccessful substitution

Fig. 3.1 Modeling of substitutions with the Adbudg function

$$t_{infl} = \left[\frac{d_i(c_i - 1)}{c_i + 1} \right]^{\frac{1}{c_i}}.$$

Finally, $f_i(t)$ is constant if $a_i = b_i$ and it is concave if

$$\frac{d_i(c_i - 1)}{(c_i + 1)} < 0.$$

Forecast

The analysis assumes that the forecast for the usage share of new product by customer i at time $T+k$, $\widehat{z}_i$, is equal to the value of the migration profile at time $T + k$, i.e.

$$\widehat{z}_i = f_i(T + k).$$

Construction of a confidence interval for the forecast $\widehat{z}_i$ requires knowledge of its probability distribution. Often it is assumed that $\widehat{z}_i$ is normally distributed. This is not the case, since there is heterogeneity across all customers, and there is no consensus on what distribution is reasonable.

The probability distribution can be estimated via bootstrapping algorithm [Efron (1979)]. The main idea of the bootstrapping algorithm, is that activity data sample $\{(z_{i1}, 1), \ldots, (z_{iT}, T)\}$ is treated as if the data points in this sample is a whole population. Re-sampling $T_1 < T$ data points at random from the proxy population $\{(z_{i1}, 1), \ldots, (z_{iT}, T)\}$ creates

$$L < \frac{T!}{T_1!(T - T_1)!}$$

data sets $z_i^{(l)}$ for $l = 1, \ldots, L$, and for each $z_i^{(l)}$, migration profile $f_i^{(l)}(t)$ yields forecast $\widehat{z}_i^{(l)}$. Then the mean μ_i and the standard deviation σ_i of $\widehat{z}_i$ are estimated by

$$\mu_i \approx \overline{z}_i = \frac{1}{L}\sum_{l=1}^{L} \widehat{z}_i^{(l)}, \qquad \sigma_i \approx \frac{1}{L-1}\sum_{l=1}^{L}\left(\widehat{z}_i^{(l)} - \overline{z}_i\right)^2.$$

Finally, Chebyshev's inequality

$$P(|X - \mu_{iT}| \geq k\sigma_i) \leq \frac{1}{k^2},$$

with k being a constant, implies that the conservative α-confidence interval for $\widehat{z}_i$ is

$$\left(\mu_i - \frac{\sigma_i}{\sqrt{1-\alpha}}, \mu_i + \frac{\sigma_i}{\sqrt{1-\alpha}}\right).$$

Segmentation

Shapes of the Adbudg function suggest the following seven customers' segments:

(1) *New customers* use only the new substitute and have never used the legacy product (see Figure 3.2 (a)). This group corresponds to the constant migration profile $f_i(t)$ with $a_i = b_i = 1$.
(2) *Balanced customers* use both products; and usage share of the new substitute does not change throughout the entire period of observation (see Figure 3.2 (b)). This group corresponds to the constant migration profile $f_i(t)$ with $a_i = b_i < 1$.
(3) *Early adopters* use mostly the new substitute; and usage of the legacy product constantly declines, (see Figure 3.2 (c)). This group has an increasing S-shape migration profile $f_i(t)$ with $(a_i - b_i)c_i d_i > 0$, $d(c-1)/(c+1) \geq 0$ and $1 \leq t_{infl} \leq T'$, where T' is a predetermined threshold.
(4) *Late adopters* are similar to the early adopters but are still in the migration process (see Figure 3.2 (e)). This group has an increasing S-shape migration profile $f_i(t)$ with $(a_i - b_i)c_i d_i > 0$, $d(c-1)/(c+1) \geq 0$ and $t_{infl} \geq T'$.
(5) *Returning customers* use mostly the legacy product; and usage of the new substitute continuously decreases (see Figure 3.2 (d)). This group has a decreasing S-shape migration profile $f_i(t)$ with $(a_i - b_i)c_i d_i < 0$ and $d(c-1)/(c+1) \geq 0$.
(6) *Conservative customers* use mostly the legacy product (see Figure 3.2 (f)).
(7) *Others* are not in any of the above segments.

This segmentation provides an insight on customers' migration patterns and helps companies to target customers for marketing campaigns.

3.3 Case Study

This section presents a case study that illustrates the customers' migration analysis. The case study uses monthly data for substitute service usage shares of 1000 customers ($m = 1000$) over a 13-month period. The first

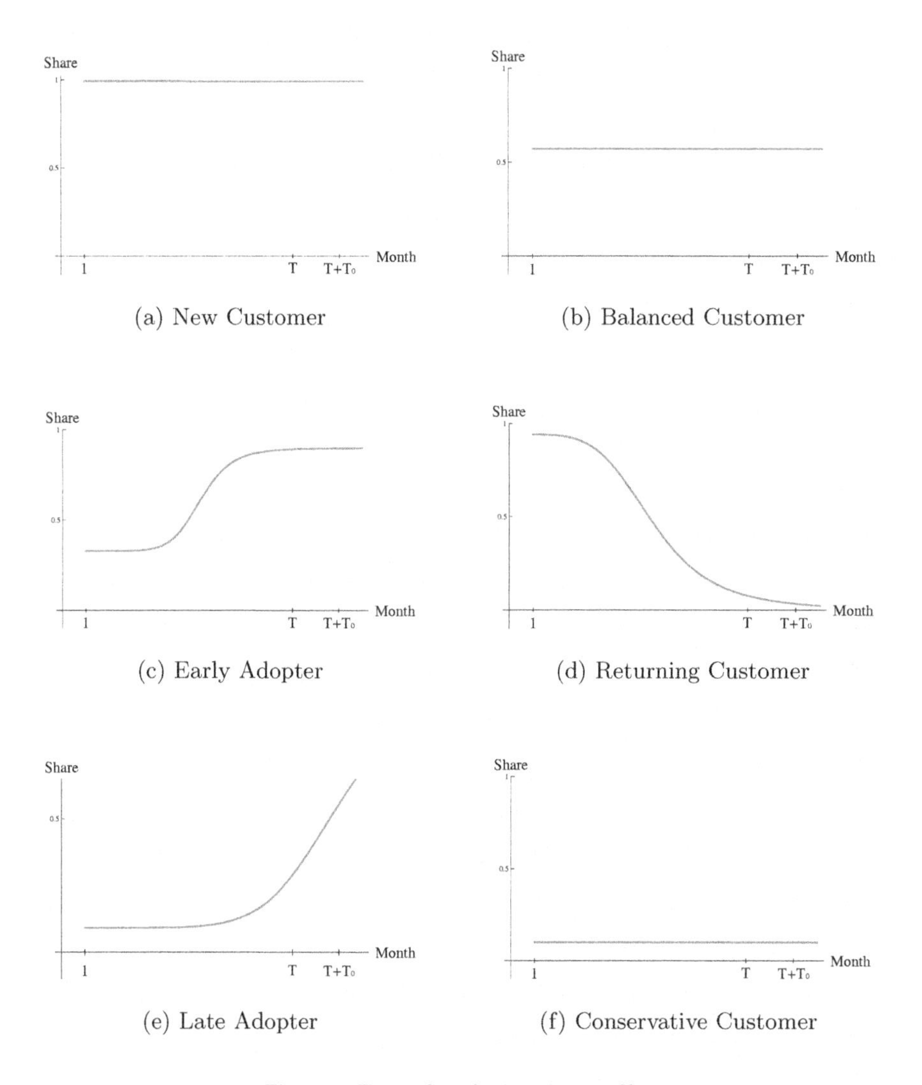

Fig. 3.2 Examples of migration profiles

ten months of the data are used for analysis ($T = 10$) and the remaining three months are used for testing ($k = 3$). Optimal values a_i, b_i, c_i and d_i in (5.2) for each customer $i = 1, \ldots, 1000$ are obtained in Mathematica. Figure 3.3 shows activity data and the migration profile of a randomly chosen customer (let it be customer 1, $i = 1$), and it also shows the three-month-ahead forecast for the substitute service usage share for 13-th month, i.e. $\widehat{z}_1 = f_1(13) = 0.8661$. The bootstrapping algorithm is implemented in

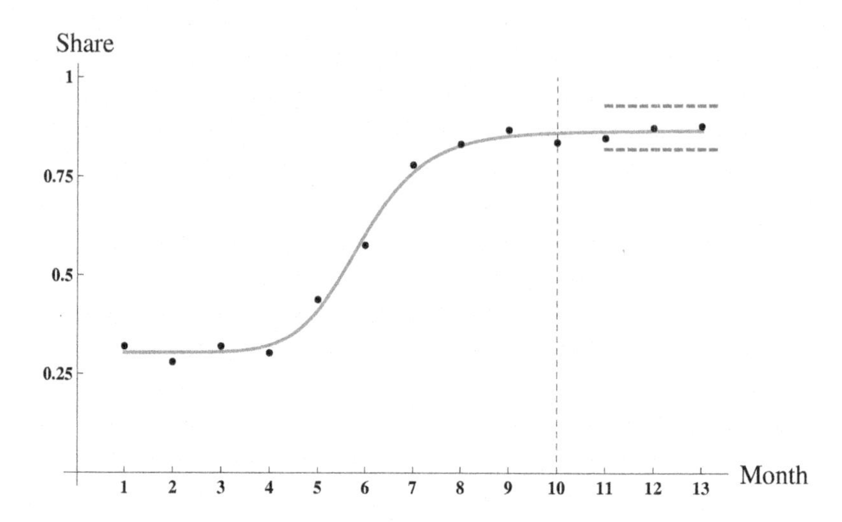

Fig. 3.3 Illustration of the analysis for a random customer

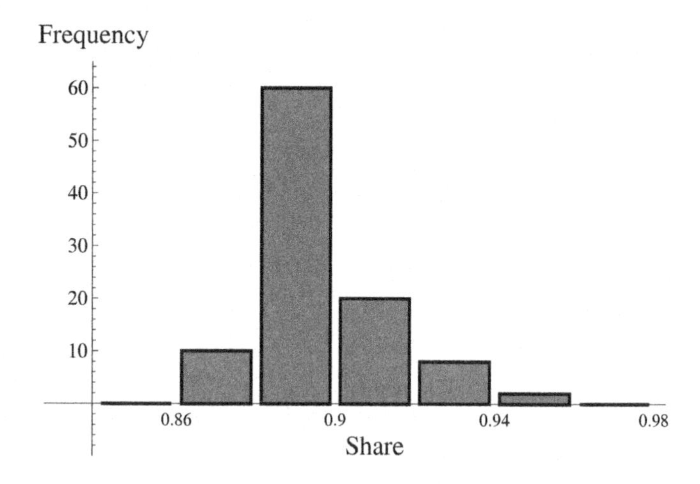

Fig. 3.4 Histogram of the distribution of the substitute share for a random customer

Mathematica for $T_1 = 7$ and $L = 100$. Figure 3.4 and Figure 3.3 show the estimated probability distribution and the 95% confidence interval for the forecast $\widehat{z}_1$, respectively. The actual value of the substitute service usage share for the 13-th month, $z_{1\ 13} = 0.8626$ lies within the 95% confidence interval. The absolute error of the forecast for the customer is equal to 0.0035. The same procedure is applied for all customers: in 87% of all

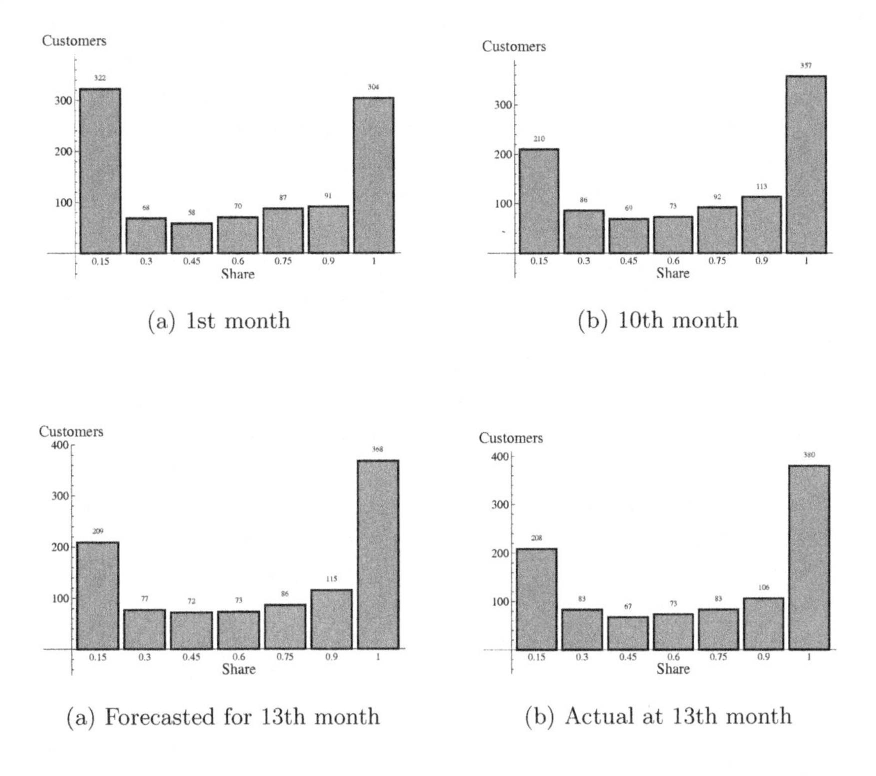

(a) 1st month

(b) 10th month

(a) Forecasted for 13th month

(b) Actual at 13th month

Fig. 3.5 Dynamics of the probability distribution of the substitute service' usage share in the entire customer sample

cases $z_{i\,13}$-s lie within the forecasted 95% confidence interval, and the mean absolute error of the forecasts for the entire customer sample is equal to 0.039. These results suggest that the obtained forecasts are reasonably accurate.

Figures 3.5 (a) and (b) show the distribution of the substitute service usage share z_{it} for $t = 1$ and $t = 10$, respectively for $i = 1, \ldots, 1000$ and illustrate migration of customers from the legacy service to its substitute. Figure 3.5 (c) shows the estimate for probability distribution of the forecast $\widehat{z}_i$ for $i = 1, \ldots, 1000$ and suggests that customer' migration from the legacy service to the substitute will continue for the next three months. Figure 3.5 (d) shows actual empirical distribution of the substitute service usage share z_{it} for $t = 13$ and validates the suggestion. Table 3.1 presents percentage of customers in each segment, and Figure 3.6 presents examples of customers' profiles for each segment. The analysis shows that the majority

Table 3.1 Segmentation summary

Segment	%
New	23
Balanced	21
Early Adopters	20
Late Adopters	5
Returning	4
Conservative	26
Other	1

of customers have migrated to the new substitute or are still in the process of migration. The company should expand its infrastructure to satisfy customer's demand for the new substitute service. Balanced and conservative customer segments are the most appropriate candidates for the targeted retention campaigns.

3.4 Practical Implications and Conclusions

Nowadays, companies face migration of customers from their legacy products and services to new technological substitutes on a daily basis. Solving this problem requires a number of tactical decisions. It is not enough to analyze just aggregate characteristics of customer migration, it also needs detailed analysis of migration patterns of each individual customer. The chapter analyzes historical activity data for each individual customer and forecasts customers' migration behavior. Customers' migration profiles are segmented according to their migration patterns. The following segments are identified: new customers, conservative customers, early adopters, late adopters, returning customers, balanced customers and customers with mixed behavior.

To illustrate the segments, consider wireless telephony as a substitute for fixed-line telephony. Wireless telephony has become an integral part of everyday life all over the world. However, customers have different patterns of migration to this technology. For example, young adults with high

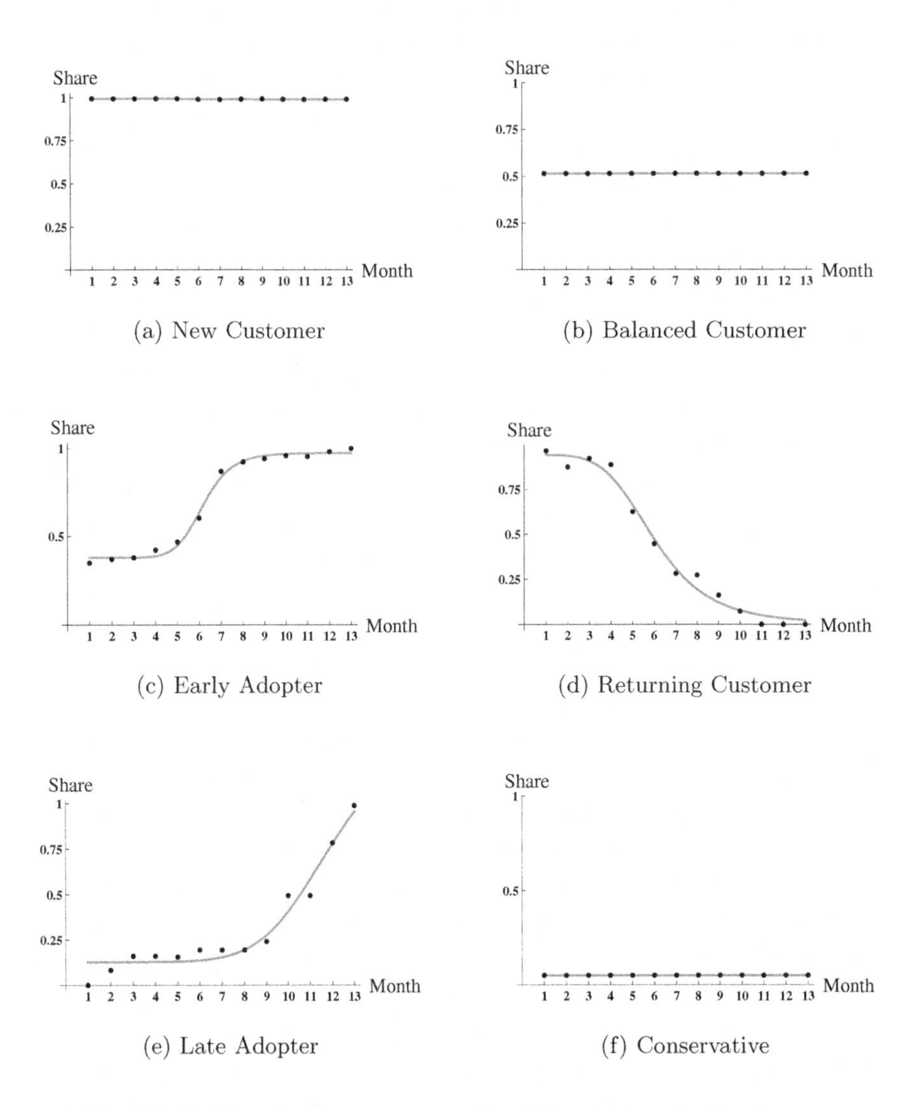

(a) New Customer

(b) Balanced Customer

(c) Early Adopter

(d) Returning Customer

(e) Late Adopter

(f) Conservative

Fig. 3.6 Profiles of customers, representing different segments of customers

income and constituting segment of the early adopters, have migrated to the wireless telephony once it emerged. They have been followed by late adopters- adults who sought convenience in business and personal communication. Nowadays, most of the adult adopters of the wireless telephony keep their fixed-line phone subscription and thus belong to the balanced customer segment. Senior customers are used to fixed-line telephony and are not interested in new technologies. This group of customers belongs to the conservative customer segment. On the other hand, today's high school and college students have grown up with wireless technology firmly integrated in their day-to-day experience. They use mostly wireless telephony and belong to the new customers' segment.

Managers can use the analysis presented in this chapter to make decisions on adoption of new technologies, infrastructure planning as well as on targeting customers for marketing campaigns. For example, if the segment of returning customers is the largest, technological substitute is no threat to the legacy product. However if the majority of customers is early or late adopters, the company's infrastructure should be upgraded to satisfy customers' demand for new technologies. The segments of conservative and balanced customers are the most appropriate candidates to be targeted for retention campaigns.

Chapter 4

Optimal Service Pricing (Single-Period Model)

4.1 Overview

Service pricing is one of the most important components of marketing strategy [Rao (1984); Nagle (1987); Marn and E. V. Roegner (2003)]: setting a price for a service too high reduces number of prospective subscribers, while setting a price too low undermines the service's perceived value. Well-known approaches to price setting include cost-plus, return-on-investment (ROI), and perceived value pricing. The cost-plus approach sets service's price to cover all costs associated with the service [Hanson (2006)], while the ROI approach sets prices to achieve a targeted return on investment [Pride *et al.* (2008)]. The perceived value pricing approach, being most challenging of the three, sets the service price based on customers' perceptions of the service's value (customers are surveyed for maximal prices that they are willing to pay for the service) [Breidert (2006)].

This chapter considers two optimal single-period service pricing models. The first model determines an optimal service subscription fee that maximizes the total profit from a service on a monopolistic market. The second model determines a Nash equilibrium for subscription fees maximizing profits from two competitive and substitutable services on a duopolistic market. Both models assume fixed service costs and account for customers' willingness to pay for the subscriptions.

4.2 Optimal Pricing in Monopoly (Single-Period Model)

This section finds an optimal service subscription fee that maximizes the total profit from a service on a monopolistic market. Let M be the total number of potential service subscribers. Assume that demand for the service is determined by continuously distributed customers' reservation prices (maximal fees that customers are willing to pay for the service subscription) with a probability density function $f(r)$, $r \geq 0$. Customers subscribe to the service, if their reservation prices are above the actual service subscription fee. Thus, a fraction of the potential market that will subscribe to the service with subcsription fee p is equal to

$$F(p) = \int_{p}^{\infty} f(r)\, dr, \tag{4.1}$$

see the area $F(p)$ on the Figure 4.1. The total profit from the service is the following:

$$\Pi(p) = M(p - c)\, F(p), \tag{4.2}$$

where c is a cost of the service.

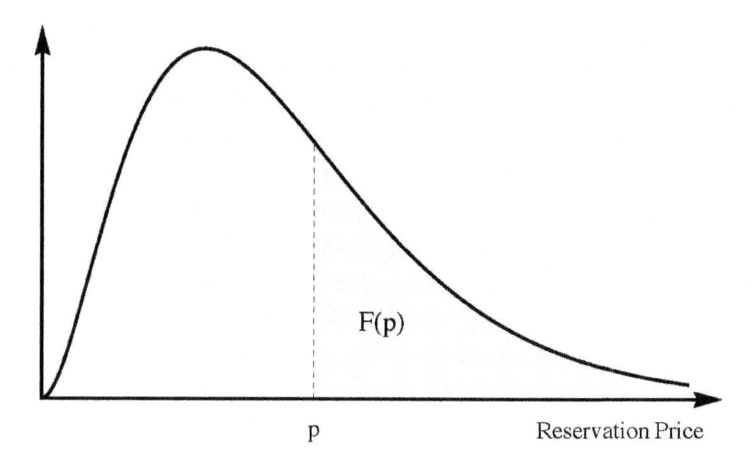

Fig. 4.1 Probability distribution of reservation prices

The necessary first order optimality condition for the optimal subscription fee p^*, maximizing the total profit from the service, is

$$\frac{d\Pi(p)}{dp} = M\left(\int_{p}^{\infty} f(r)\, dr - (p - c)f(p)\right) = 0, \tag{4.3}$$

and the second order sufficient condition is

$$\frac{d^2\Pi(p)}{d^2p} = -M(2f(p) + (p - c)f'(p)) < 0. \tag{4.4}$$

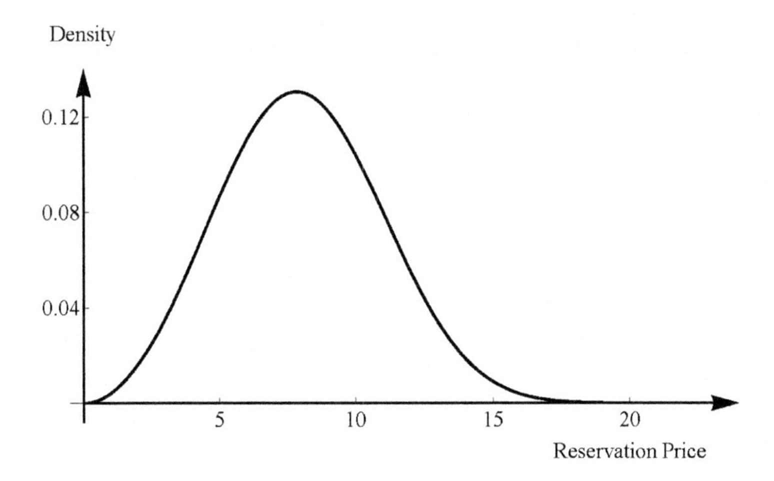

Fig. 4.2 Weibull probability density function with parameters $a = 3$ and $b = 9$

As an example, let the potential market have 100 customers, i.e., $M = 100$, and let customers' reservation prices follow the Weibull probability distribution with a shape parameter $a = 3$ and a scale parameter $b = 9$:

$$f(r) = ab^{-a}\,\mathrm{e}^{-(r/b)^a}\,r^{a-1}, \quad r \geq 0, \tag{4.5}$$

see Figure 4.2 (on average customers are willing to pay for the subscription about \$8). For the fixed cost $c = \$\,1$, equation (4.3) implies that the optimal subscription fee is $p^* = \$\,6.98315$ (below the average fee that customers are willing to pay for the subscription). Figures 4.3 and 4.4 show how the optimal subscription fee p^* and the optimal total profit $\Pi(p^*)$ depend on the cost c.

4.3 Optimal Pricing in Duopoly (Single-Period Model)

Let two companies A and B provide substitutable services A and B respectively. Assume that demand for the services is determined by customers' reservation prices r_1 and r_2 with a joint probability density function $f(r_1, r_2)$. The goal of this section is to find a Nash equilibrium (p_1^*, p_2^*) for subscription fees for the services A and B, such that both service providers exceed their maximal profits (in other words, any deviation from the Nash equilibrium is unprofitable for both service providers).

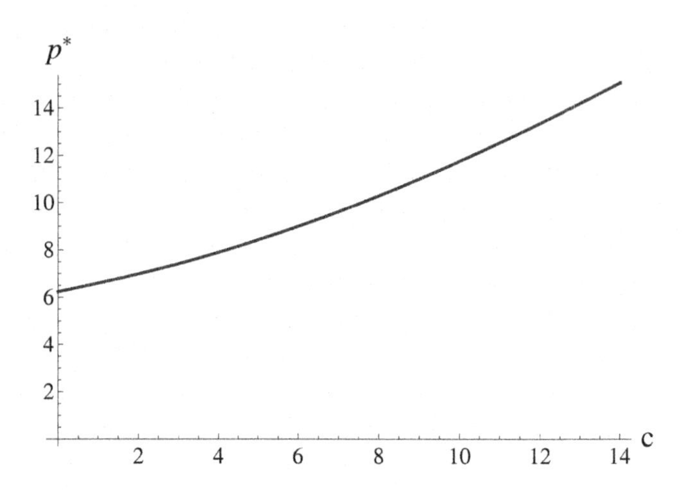

Fig. 4.3 Optimal subscription fee vs. cost

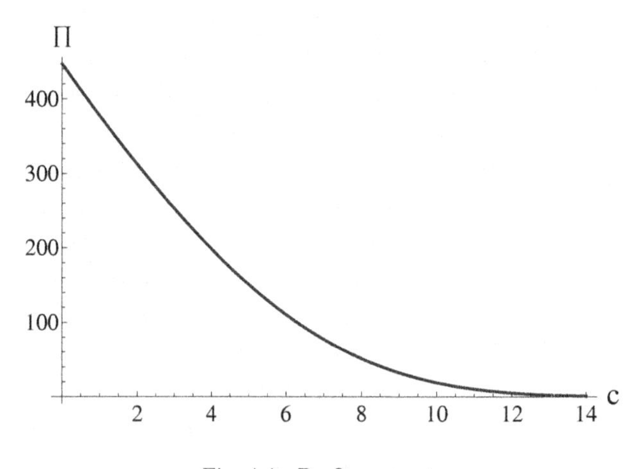

Fig. 4.4 Profit vs. cost

Customers choose a service that maximizes a surplus from subscription (difference between the reservation price and the actual service subscription fee). A customer subscribes to service A if $r_1 > p_1$ and

$$r_1 - p_1 > r_2 - p_2,$$

i.e. a surplus from subscription to service A is greater than that from subscription to service B. Similarly, a customer subscribes to service B if $r_2 > p_2$ and

$$r_2 - p_2 \geq r_1 - p_1.$$

A customer does not subscribe to any service, if $p_1 > r_1$ and $p_2 > r_2$. These conditions can be represented in the form of sets:

$$F_0 = \{(r_1, r_2) | p_1(t) > r_1 \text{ and } p_2(t) > r_2\},$$
$$F_1 = \{(r_1, r_2) | r_1 > p_1(t) \text{ and } r_1 - p_1(t) > r_2 - p_2(t)\},$$
$$F_2 = \{(r_1, r_2) | r_2 > p_2(t) \text{ and } r_2 - p_2(t) \geq r_1 - p_1(t)\},$$

see Figure 4.5.

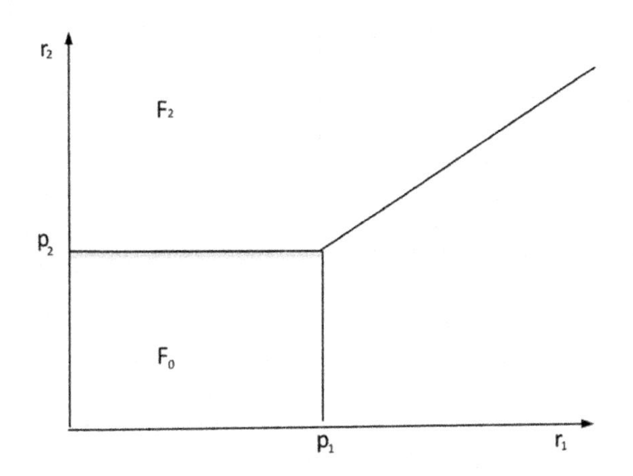

Fig. 4.5 Customer's preferences on the duopolistic market

If c_1 and c_2 are costs of services A and B respectively, then the total profit from service A is given by

$$\Pi_1(p_1, p_2) = M(p_1 - c_1)\left(\int_{p_1}^{\infty}\int_{0}^{p_2} f(r_1, r_2)dr_2dr_1 + \int_{p_2}^{\infty}\int_{p_1+r_2-p_2}^{\infty} f(r_1, r_2)dr_1dr_2\right), \tag{4.6}$$

and similarly, the profit from service B takes the form

$$\Pi_2(p_1, p_2) = M(p_2 - c_2)\left(\int_{p_2}^{\infty}\int_{0}^{p_1} f(r_1, r_2)dr_1dr_2 + \int_{p_1}^{\infty}\int_{p_2+r_1-p_1}^{\infty} f(r_1, r_2)dr_2dr_1\right), \tag{4.7}$$

where M is a market size. The pair of prices (p_1^*, p_2^*) is a Nash equilibrium if

$$\Pi_1(p_1, p_2^*) \leq \Pi_1(p_1^*, p_2^*) \quad \forall p_1, \tag{4.8}$$

and

$$\Pi_2(p_1^*, p_2) \leq \Pi_2(p_1^*, p_2^*) \quad \forall p_2. \tag{4.9}$$

The inequalities (4.8) and (4.9) imply

$$\left.\frac{\partial \Pi_1(p_1,p_2)}{\partial p_1}\right|_{(p_1^*,p_2^*)} = 0, \qquad \left.\frac{\partial \Pi_2(p_1,p_2)}{\partial p_2}\right|_{(p_1^*,p_2^*)} = 0. \tag{4.10}$$

With (4.6) and (4.7), (4.10) yields the first-order necessary conditions:

$$\begin{aligned}
&\int_{p_1}^{\infty}\int_{0}^{p_2} f(r_1,r_2)dr_2dr_1 + \int_{p_2}^{\infty}\int_{p_1+r_2-p_2}^{\infty} f(r_1,r_2)dr_1dr_2 - \\
&\qquad (p_1-c_1)\left(\int_0^{p_2} f(p_1,r_2)dr_2 + \int_{p_2}^{\infty} f(p_1+r_2-p_2,r_2)dr_2\right) = 0, \\
&\int_{p_2}^{\infty}\int_{0}^{p_1} f(r_1,r_2)dr_1dr_2 + \int_{p_1}^{\infty}\int_{p_1+r_2-p_2}^{\infty} f(r_1,r_2)dr_2dr_1 - \\
&\qquad (p_2-c_2)\left(\int_0^{p_1} f(r_1,p_2)dr_1 + \int_{p_1}^{\infty} f(r_1,p_2+k(r_1-p_1))dr_1\right) = 0.
\end{aligned} \tag{4.11}$$

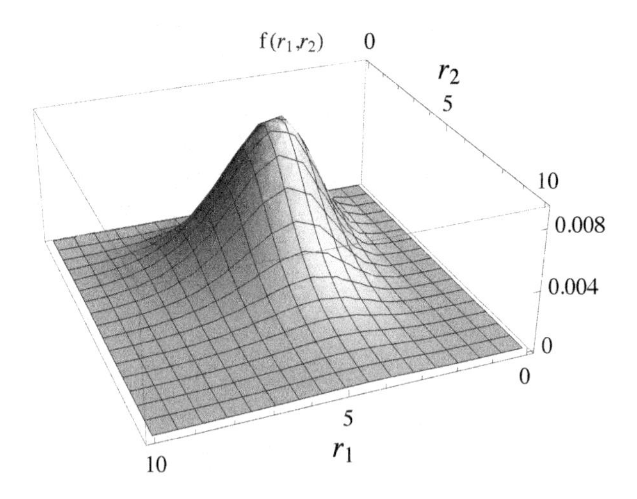

Fig. 4.6 Bivariate gamma joint distribution of reservation prices

For illustration, consider the following example. Assume that customers' reservation prices r_1, r_2 have the bivariate gamma joint probability density [D'Este (1981)]:

$$f(r_1, r_2) = \left(1 + \lambda \prod_{j=1}^{2} (2G(r_j, \alpha_j) - 1)\right) \prod_{j=1}^{2} \frac{r_j^{\alpha_j - 1} e^{-r_j}}{\Gamma(\alpha_j)}, \tag{4.12}$$

where $\alpha_j \geq 0$, $r_j > 0, j = 1, 2$; $|\lambda| \leq 1$ and $G(r_j, a_j)$ is a cumulative density function of a random variable with a probability density function $g(r_j, \alpha_j) = e^{-r_j} r_j^{\alpha_j - 1}/\Gamma(\alpha_j)$.

For illustration, let $\lambda = 1$, $\alpha_1 = 5$, and $\alpha_2 = 3$ (see Figure 4.6), and let $c_1 = \$\,2$ and $c_2 = \$\,1$ (the costs of the services A and B, respectively). Substitution of (4.12) into (4.11) yields the Nash equilibrium for the subscription fees $(p_1^*, p_2^*) = (\$\,3.95064, \$\,2.92025)$.

Figure 4.7 shows cross-sections of surfaces $\partial\Pi_1(p_1, p_2)/\partial p_1$ and $\partial\Pi_2(p_1, p_2)/\partial p_1$ in the plane $Z(p_1, p_2) = 0$, which intersect at (p_1^*, p_2^*).

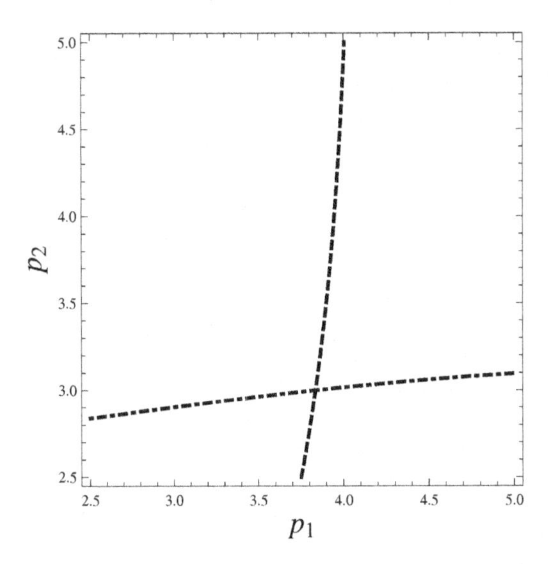

Fig. 4.7 Cross-sections of surfaces $\frac{\partial\Pi_1(p_1,p_2)}{\partial p_1}$, $\frac{\partial\Pi_2(p_1,p_2)}{\partial p_2}$ in the plane $Z(p_1, p_2) = 0$

Managers can use models presented in this chapter for developing optimal service pricing strategies. Distribution of customers' reservation prices can be determined by customer surveying.

Chapter 5

Optimal New Service Pricing in Monopoly (Dynamic Model)

5.1 Overview

Pricing of telecommunication services has been and continues to be one of leading research areas in marketing. Almost all US households subscribe to at least one telecommunication service [US Census Bureau (2009)]: fixed-line or wireless phone, cable or satellite television, DSL, cable or broadband internet, in-store or by-mail DVD rental, etc. Moreover, constant advances in the telecommunication industry result in an extensive growth of new subscription services. What affects price of a new or existing telecommunication service and how to correctly price the service? Answering these questions is the subject of this work. Setting price of a new service too low decreases potential revenues of a provider and undermines the service's market value, while setting prices too high discourages potential service adopters. Successful pricing decisions require accurate projections for future demand and diffusion of the new service, which depend on several interdependent factors: service advertising, word-of-mouth effect, customer churn, willingness to pay, etc.

Existing optimal pricing strategies suggest that the price of a durable good should decrease in time, whereas the subscription fee of a new service should increase. The rationale is the following. A producer of a durable good charges a high price at the introductory stage targeting the high-end customers, who are willing to pay the high price, and then reduces the price to reach the next segment of adopters [Rohlfs (1974)]. On the other hand, a provider of a new service should set the initial price low enough to encourage new subscriptions and then should raise the price gradually over the time [Rohlfs (1974)]. However, existing optimal service pricing strategies deal only with an optimal subscription fee rather than both: an optimal subscription fee and an optimal activation/installation fee. This work develops a new service pricing model and based on this model, it derives an optimal service pricing strategy for an optimal subscription fee and a one time activation fee. The optimal fees vary with time and maximize a total profit of a service provider.

The chapter is organized into four sections. Section 5.2 develops the new service pricing model and states a theorem on an optimal service pricing strategy. Section 5.3 illustrates the optimal service pricing strategy numerically. Section 5.4 discusses limitations of the model and concludes the chapter.

5.2 Model and Optimal Pricing Strategy

Let M be the total number of potential adopters for a monopolist service provider. Each new customer, subscribing at time t, is charged a one-time activation fee $q(t)$, and then he/she is charged a periodical monthly subscription fee $p(t)$. The customer has a choice to discontinue the service at any time at no cost. However, if the customer discontinues the service and then decides to subscribe again, he/she is considered as a new customer and is required to pay the activation fee again.

Let $x(t)$ be the number of service subscribers at time t with $\dot{x}(t) = dx/dt$ representing the rate of change in the number of subscribers at time t. As in the Bass [Bass (1969)] model, we assume that the number of customers $f(t)$ adopting the service at time t is proportional to the number of potential adopters $M - x(t)$ and to the number of current customers $a + bx(t)$ with a and b representing advertising and word-of-mouth effects, respectively. However, we also assume that customers adopt the new service if their reservation prices for subscription and activation are higher than the actual subscription and activation fees. In other words, $f(t)$ is assumed to be proportional to the fraction of customers who are willing to pay the activation fee $q(t)$ and subscription fee $p(t)$. Suppose the activation and subscription reservation fees are independently exponentially distributed with the expected values to be $1/\alpha$ and $1/\beta$, respectively. Then $\mathbb{P}\{s_1 \geq p(t),\, s_2 \geq q(t)\} = \mathbb{P}\{s_1 \geq p(t)\}\mathbb{P}\{s_2 \geq q(t)\} = \mathrm{e}^{-\alpha p(t)}\,\mathrm{e}^{-\beta q(t)}$ is the fraction of potential adopters willing to pay the activation fee $q(t)$ and subscription fee $p(t)$. Consequently, the number of customers who adopt the service at time t is given by

$$f(x(t), p(t), q(t)) = (M - x(t))(a + bx(t))\,\mathrm{e}^{-\alpha p(t) - \beta q(t)}\,. \tag{5.1}$$

Under assumption that the customers discontinue the service at a constant (churn) rate γ, the diffusion of the new service demand is modeled by

$$\dot{x}(t) = f(x(t), p(t), q(t)) - \gamma\, x(t). \tag{5.2}$$

A service provider's goal is to maximizes the total profit over the planning horizon $[0, T]$:

$$\Pi(x, \dot{x}, p, q) = \int_0^T x(t)(p(t) - c_1) + f(x(t), p(t), q(t))(q(t) - c_2) dt \tag{5.3}$$

subject to (5.1) and (5.2), where constants $c_1 \geq 0$ and $c_2 \geq 0$ are costs associated with subscription (maintenance) and activation (installation), respectively. Figure 5.1 shows finding an optimal pricing policy $\{p^*(t), q^*(t)\}$ as an optimal control problem.

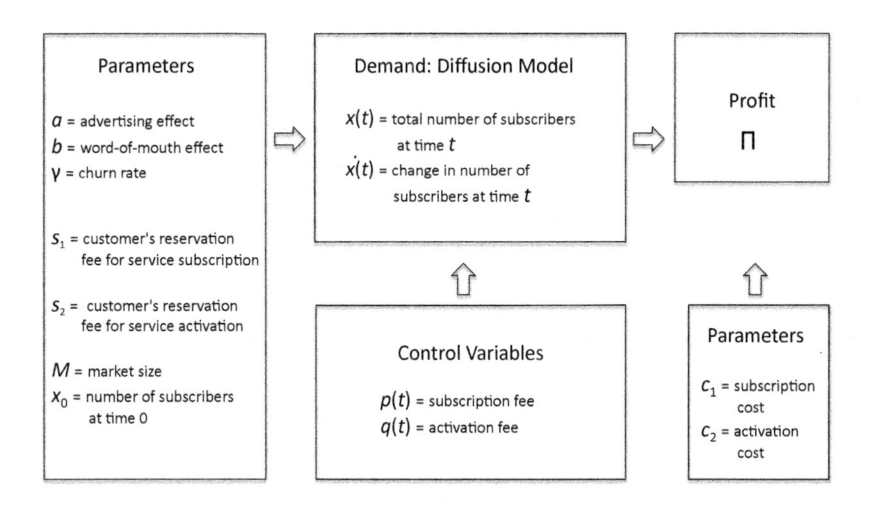

Fig. 5.1 Optimal service pricing model

Theorem 5.1 (optimal pricing strategy). *Let* (5.1) *and* (5.2) *model diffusion of a new service. Then one-time activation fee* $q^*(t)$ *and subscription fee* $p^*(t)$ *that maximize* (5.3) *are determined by*

$$\begin{aligned} q^*(t) =& \frac{1}{\gamma\,\alpha - \beta}\Big(\left(g(T) - c_1\alpha - c_2\beta\right) \mathrm{e}^{(t-T)(\gamma-\beta/\alpha)} - g(t) \\ &+ \alpha(c_1 + \gamma\,(c_2 + 1/\beta)) - 1\Big), \\ p^*(t) =& \frac{1}{\alpha}\left(\ln\left[\frac{\alpha}{x_0\beta}\left(M\,\mathrm{e}^{(\gamma-\beta/\alpha)t} - x_0\right)\left(a + b\,x_0\,\mathrm{e}^{(\beta/\alpha-\gamma)t}\right)\right] - \beta\,q^*(t)\right), \end{aligned} \tag{5.4}$$

where

$$g(t) = \ln\left[\frac{1}{x_0\beta}\left(a\left(M\,\mathrm{e}^{(\gamma-\beta/\alpha)t} - x_0\right) + bx_0(M - x_0\,\mathrm{e}^{(\beta/\alpha-\gamma)t})\right)\right].$$

Proof.

Optimal pricing policy $\{p^*(t), q^*(t)\}$ is found from the Calculus of Variations problem with the variable endpoint for $x(t)$ at $t = T$:

$$\begin{aligned} &\max_{p(t),\,q(t)} \;\Pi(x, \dot{x}, p, q) \\ &\text{s.t.}\quad \dot{x}(t) = f(x(t), p(t), q(t)) - \gamma\, x(t), \quad x(0) = x_0, \quad t \in [0, T]. \end{aligned} \tag{5.5}$$

This problem can be solved by the Lagrange multipliers technique [Gelfand and Fomin (1963); Giaquinta and Hildebrandt (1996)] or by applying Pontryagin maximum principle [Pontryagin (1962); Geering (2007)]. Problem solutions using both techniques are presented below.

First, let us apply the Lagrange multipliers technique. The Lagrangian for (5.5) is given by

$$\mathcal{F}(x, \dot{x}, p, q) = \int_0^T \mathcal{L}(x(t), \dot{x}(t), p(t), q(t), \lambda(t))dt,$$

where

$$\begin{aligned} \mathcal{L}(x, \dot{x}, p, q, \lambda) = x(t)(p(t) - c_1) + (\dot{x}(t) + \gamma\, x(t))(q(t) - c_2) + \\ + \lambda(t)\left(\dot{x}(t) - f(x(t), p(t), q(t)) + \gamma\, x(t)\right), \end{aligned} \tag{5.6}$$

and $\lambda(t)$ is the Lagrange multiplier, representing a net benefit of acquiring an additional customer at time t. The total variation of the functional $\mathcal{F}$ takes the form

$$\begin{aligned} \delta\mathcal{F} = \int_0^T \left[\left(\frac{\partial\mathcal{L}}{\partial x} - \frac{d}{dt}\left(\frac{\partial\mathcal{L}}{\partial\dot{x}}\right)\right)\delta x + \frac{\partial\mathcal{L}}{\partial p}\delta p + \frac{\partial\mathcal{L}}{\partial q}\delta q + \frac{\partial\mathcal{L}}{\partial\lambda}\delta\lambda\right] dt \\ + \left[(q - c_2 + \lambda)\,\delta x\right]\Big|_0^T, \end{aligned}$$

and the necessary (first-order) optimality condition for the functional $\mathcal{F}$ to have an extremum, i.e., $\delta\mathcal{F} = 0$, reduces to the Euler-Lagrange equations:

$$\begin{aligned} \frac{\partial\mathcal{L}}{\partial x} - \frac{d}{dt}\left(\frac{\partial\mathcal{L}}{\partial\dot{x}}\right) = p(t) - c_1 + \lambda(t)(a - bM + 2bx(t))\,\mathrm{e}^{-\alpha p(t)-\beta q(t)} \\ + \gamma(q(t) - c_2 + \lambda(t)) - \dot{q}(t) - \dot{\lambda}(t) = 0, \end{aligned} \tag{5.7a}$$

$$\frac{\partial\mathcal{L}}{\partial p} = x(t) + \alpha\,\lambda(t)(M - x)(a + bx)\;\mathrm{e}^{-\alpha p(t)-\beta q(t)} = 0, \tag{5.7b}$$

$$\frac{\partial\mathcal{L}}{\partial q} = \dot{x}(t) + \gamma\, x(t) + \beta\,\lambda(t)\,(M - x(t))(a + bx(t))\,\mathrm{e}^{-\alpha p(t)-\beta q(t)} = 0, \tag{5.7c}$$

$$\frac{\partial\mathcal{L}}{\partial\lambda} = \dot{x}(t) + \gamma\, x(t) - (M - x(t))(a + bx(t))\,\mathrm{e}^{-\alpha p(t)-\beta q(t)} = 0, \tag{5.7d}$$

with the initial condition $x(0) = x_0$ and the transversality condition

$$q(T) - c_2 + \lambda(T) = 0, \tag{5.8}$$

because $\delta x(T) \neq 0$.

Subtracting (5.7c) from (5.7d), we have $\lambda(t) = -1/\beta$ provided that $x(t) \neq M$ and $a + bx(t) \neq 0$. With this relationship, (5.7b) and (5.7d) imply that $\dot{x}(t) + (\gamma - \beta/\alpha)x(t) = 0$, $x(0) = x_0$, whence $x(t) = x_0\, \mathrm{e}^{(\beta/\alpha-\gamma)t}$. Now it follows from (5.7b) that

$$p(t) = \frac{1}{\alpha}\left(-\beta\, q(t) + \ln\left[\frac{\alpha}{x_0\beta}\left(M\,\mathrm{e}^{(\gamma-\beta/\alpha)t} - x_0\right)\left(a + b\,x_0\,\mathrm{e}^{(\beta/\alpha-\gamma)t}\right)\right]\right). \tag{5.9}$$

Finally, with (5.9), (5.7a) reduces to the first order ordinary differential equation with respect to $q(t)$ satisfying the condition $q(T) = c_2 + 1/\beta$ which follows from (5.8).

Let us apply Pontryagin maximum principle to solve the problem (5.5). The Hamiltonian for (5.5) is given by

$$\begin{aligned}\mathcal{H}(x, p, q, \mu) = (p(t) - c_1)x + (q(t) - c_2)(M - x(t))(a + bx(t))\,\mathrm{e}^{-\alpha p(t)-\beta q(t)}\\ +\mu\left((M - x(t))(a + bx(t))\,\mathrm{e}^{-\alpha p(t)-\beta q(t)} - \gamma x(t)\right),\end{aligned} \tag{5.10}$$

where $\mu(t)$ is a co-state variable representing the net benefit of acquiring of an additional customer at time t and satisfying the transversality condition at $t = T$:

$$\mu(T) = 0.$$

The optimal solution of the problem (5.5) must satisfy the following necessary (first-order) optimality conditions:

$$\begin{aligned}\frac{\partial\mathcal{H}}{\partial x} &= p(t) - c_1 - \gamma M - (a - bM + 2bM)(\mu(t) + q(t) - c_2)\,\mathrm{e}^{-\alpha p(t)-\beta q(t)}\\ &= -\dot{\mu}(t),\end{aligned} \tag{5.11a}$$

$$\frac{\partial\mathcal{H}}{\partial p} = x(t) - \alpha(M - x(t))(a + bx(t))(\mu(t) + q(t) - c_2)\,\mathrm{e}^{-\alpha p(t)-\beta q(t)} = 0, \tag{5.11b}$$

$$\frac{\partial\mathcal{H}}{\partial q} = (M - x(t))(a + bx(t))(1 - \beta(\mu(t) + q(t) - c_2))\,\mathrm{e}^{-\alpha p(t)-\beta q(t)} = 0, \tag{5.11c}$$

and the constrain equation (5.2) with the initial condition $x(0) = x_0$. From (5.11c) we obtain that

$$q(t) = 1/\beta + c_2 - \mu(t), \tag{5.12}$$

provided that $x(t) \neq M$ and $a+bx(t) \neq 0$. By substituting (5.12) in (5.11b) we obtain that

$$p(t) = \frac{1}{\alpha}\left(-\beta\, q(t) + \ln\left[\frac{\alpha}{x_0\beta}\left(M\,\mathrm{e}^{(\gamma-\beta/\alpha)t} - x_0\right)\left(a + b\,x_0\,\mathrm{e}^{(\beta/\alpha-\gamma)t}\right)\right]\right). \tag{5.13}$$

Finally, with (5.12) and (5.13), (5.11a) and (5.2) reduce to a system of the first order ordinary differential equations with respect to $\mu(t)$ and $x(t)$ satisfying conditions $\mu(T) = 0$ and $x(T) = x_0$ respectively.

The optimal solution (5.4) also satisfies the sufficient optimality conditions (second-order) derived by Arrow and Kurz [Arrow and Kurz (1970); Seierstad and Sydsaeter (1977)], since the Hamiltonian (5.10) is a concave function of the state variable $x(t)$. □

The optimal profit is determined by (5.3) with (5.4):

$$\begin{aligned}\Pi^* = &\frac{1}{\alpha\gamma-\beta}\left(x_0(1+c_1\alpha+c_2\beta)(\mathrm{e}^{-T(\gamma-\beta/\alpha)} - 1)\right.\\ &+ M\ln\left[\frac{(M - x_0\,\mathrm{e}^{-T(\gamma-\beta/\alpha)})}{M-x_0}\right]\\ &+ \frac{a}{b}\ln\left[\frac{a+bx_0}{a+bx_0\,\mathrm{e}^{-T(\gamma-\beta/\alpha)}}\right] + x_0\ln\left[\frac{\alpha(M-x_0)(a+bx_0)}{x_0\beta}\right]\\ &\left.- x_0\,\mathrm{e}^{-T(\gamma-\beta/\alpha)}\ln\left[\frac{\alpha(\mathrm{e}^{T(\gamma-\beta/\alpha)}\,M - x_0)(a+bx_0\,\mathrm{e}^{-T(\gamma-\beta/\alpha)})}{x_0\beta}\right]\right).\end{aligned} \tag{5.14}$$

Advertising and word-of-mouth inform customers about the benefits of the new service and thus increase both the awareness of the service and the customers' willingness to pay for the service [Jagpal (2008)]. Consequently, the parameters α and β can depend on a and b: $\alpha = \alpha(a,b)$ and $\beta = \beta(a,b)$. For example, $1/\alpha$ and $1/\beta$, being the expected values of the activation and subscription reservation fees, can be modeled as linear functions of a and b, i.e.,

$$\alpha(a,b) = \frac{1}{k_1a + k_2b + \alpha_0}, \qquad \beta(a,b) = \frac{1}{k_3a + k_4b + \beta_0},$$

where constants $k_1, \ldots, k_4$, α_0, and β_0 can be estimated from real data.

Section 5.3 examines dependence of the optimal service pricing policy (5.4) and the optimal profit (5.14) on the parameters of the model.

5.3 Numerical Illustration

The optimal pricing policy is illustrated in two numerical examples: Example 1 corresponds to a broadband internet service, while Example 2 models a satellite internet service.

In both examples, the market has 100 potential adopters ($M = 100$) with 10 customers subscribed to the service at $t = 0$ ($x_0 = 10$) with $T = 1$. Also, the churn rate γ and the parameters of a and b associated with advertising and word-of-mouth effects are assumed to be same: $\gamma = 0.01$ and $a = b = 0.5$. The examples differ in subscription and activation costs: $c_1 = 15$ and $c_2 = 5$ in Example 1, and $c_1 = 5$ and $c_2 = 15$ in Example 2. These costs are consistent with real data: for a broadband internet service (Example 1) as well as cell phone service, the activation fee is always less than the monthly subscription fee, while for a satellite internet service (Example 2) and satellite television service, the activation fee, including an installation charge for a satellite dish, often exceeds the monthly subscription fee. These data are summarized in Table 5.1.

Table 5.1 Values of parameters in Examples 1 and 2

Parameter	Example 1	Example 2
M	100	100
x_0	10	10
γ	0.01	0.01
T	1	1
c_1	15	5
c_2	5	15
a	0.5	0.5
b	0.5	0.5

The examples suggest different models for the expected values for the activation and subscription reservation fees:

$$\text{Example 1:} \quad 1/\alpha = 10 + 20a + 10b, \qquad 1/\beta = 5 + 10a + 5b.$$

$$\text{Example 2:} \quad 1/\alpha = 5 + 10a + 5b, \qquad 1/\beta = 20 + 30a + 20b.$$

Figures 5.2 (a)–(d) show optimal subscription and activation fees in Examples 1 and 2. The optimal pricing policies have similar patterns: the subscription fees are high at the introductory stage and decrease gradually over time (skimming pricing strategy) (Figures 5.2 (a) and (c)), and the optimal activation fees are low at the introductory stage and increase gradually over time (penetration pricing strategy) (Figures 5.2 (b) and (d)).

However, at the introductory stage, the optimal activation fee is negative in Example 1, while it is zero in Example 2. A negative activation fee can be considered as an incentive (promotion rate, rebate, etc.) for service subscription.

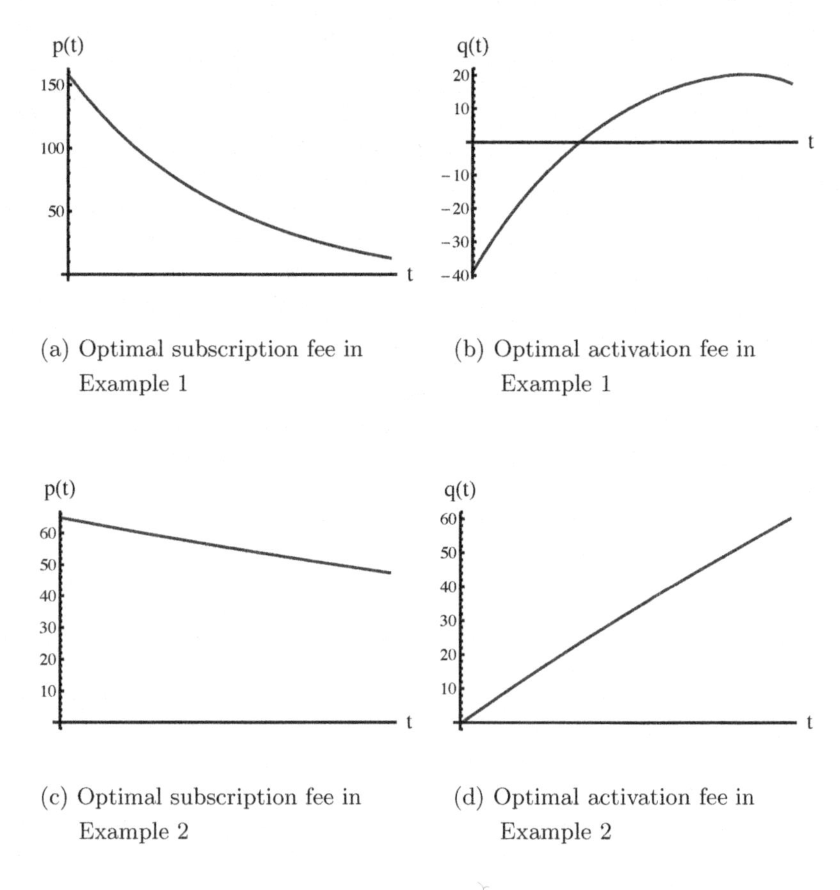

(a) Optimal subscription fee in Example 1

(b) Optimal activation fee in Example 1

(c) Optimal subscription fee in Example 2

(d) Optimal activation fee in Example 2

Fig. 5.2 Optimal pricing strategies

In both examples, the optimal service pricing strategies agree with real pricing of (broadband and satellite) internet services. For example, a broadband portal statistics collected by Organization for Economic Co-operation and Development (OECD) shows that a subscription fee of internet usage decreases over time, see [Organization for Economic Co-operation and Development (2009)]. Also, as the broadband internet service emerged, providers attract potential customers with different promotions offering them incentives such as free service during the first month, free equipment, gift cards or rebates. These promotions correspond to negative activation fees. Satellite internet companies offer free installation of a satellite dish, which corresponds to zero activation fee.

For the data in Example 1, Figures 5.3 (a) and 5.3 (b) illustrate dependence of the optimal profit on the parameters a and b; and Figures 5.3 (c) and 5.3 (d) show the optimal profit as functions of churn rate and market size parameters. This analysis agrees with the well-known fact: optimal profit increases with the parameters corresponding to advertisement, word-of-mouth effects and market size, and decreases with the churn rate.

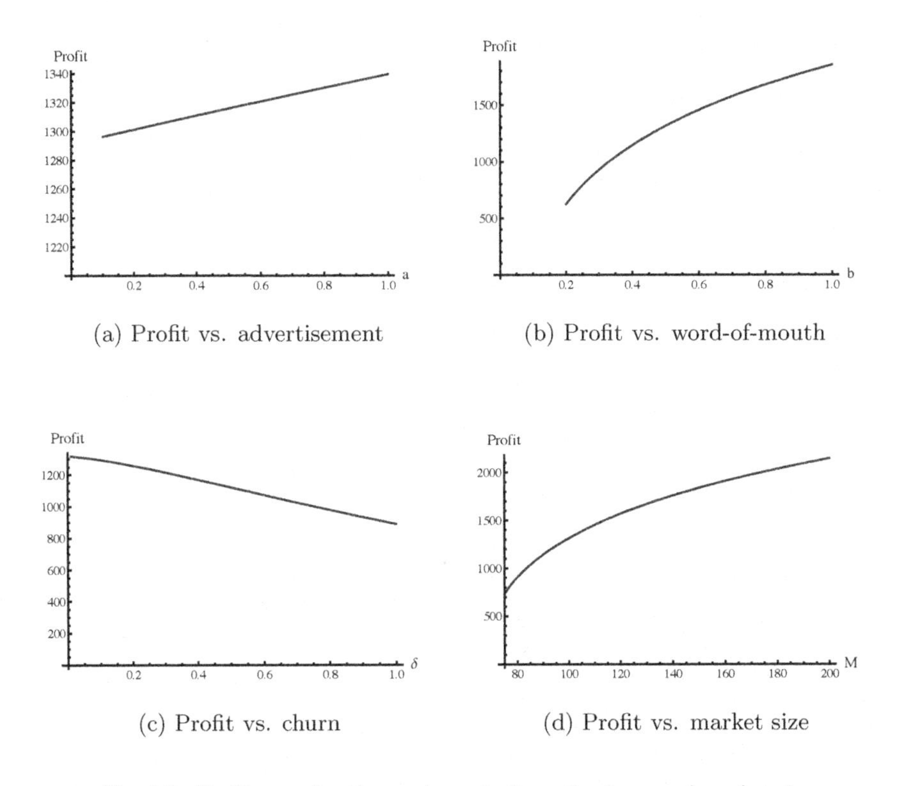

(a) Profit vs. advertisement

(b) Profit vs. word-of-mouth

(c) Profit vs. churn

(d) Profit vs. market size

Fig. 5.3 Profit vs. advertisement, word-of-mouth, churn and market size

5.4 Conclusions

This work has developed a model for dynamic optimal pricing of a subscription service. The model maximizes service provider's profit with respect to one-time activation fee and recurring subscription fee over a planning horizon. It considers advertising and word-of-mouth effects and customer churn. Numerical results suggest a penetration strategy for the activation fee and a skimming strategy for the subscription fee. These strategies are consistent with the pricing patterns observed in the telecommunication industry (broadband and satellite internet services): a subscription fee of internet usage decreases over time, whereas providers of new services attract potential customers with different promotions corresponding to a negative activation fee.

The suggested model has several limitations: (i) subscription fee is continuous; (ii) reservation fees for activation and subscription are independently distributed; (iii) distributions for reservation fees are independent of the number of potential adopters; (iv) the churn rate is constant. Addressing these limitations will provide more accurate pricing models and is the subject for the future research.

Chapter 6

Optimal New Service Pricing in Duopoly (Dynamic Model)

Service pricing strategy is one of the most important decisions directly affecting provider's profit and success. In a competitive market, especially with service substitution, a service provider should consider not only customers' willingness to pay, but also competitors' prices. It is well known that changes in prices of competitors' substitute products and service demand are intimately related [Birge *et al.* (1998)]. This chapter considers a market with two competitive service providers (duopoly) and determines optimal dynamic subscription fee strategies that maximize profits from two substitutable services over a given planning horizon. Depending on service fees and willingness to pay for each service, a customer either subscribes to a single most preferable service or subscribes to none.

6.1 Model

Suppose that two substitutable new services (service 1 and service 2) compete on the market of potential adopters with size M. Let $x_i(t)$, $i = 1, 2$ denote the number of subscribers of service i at time t with $\dot{x}_i(t) = dx_i/dt$, $i = 1, 2$ representing the rate of change in the number of subscribers of service i at time t. Each customer, subscribing to service i ($i = 1, 2$) is charged a periodical subscription fee $p_i(t)$. Assume that a customer may subscribe only to a single service at time t and may discontinue his/her service subscription at any time at no cost. Let r_1 and r_2 be maximal fees that customers are willing to pay for a one-period subscription to services 1 and 2 respectively (reservation prices) with a probability density function $w(r_1, r_2)$; and let $p_1(t)$ and $p_2(t)$ denote subscription fees for services 1 and 2 respectively. A customer subscribes to service 1 at time t if $r_1 > p_1(t)$ and $r_1 - p_1(t) > r_2 - p_2(t)$, i.e., a surplus from subscription to service 1 is greater than that from service 2. Similarly, a customer subscribes to service 2 at time t if $r_2 > p_2(t)$ and $r_2 - p_2(t) \geq r_1 - p_1(t)$. And finally, customer does not subscribe to either service, if $p_1(t) > r_1$ and $p_2(t) > r_2$. Consequently, the sets

$$F_0(t) = \{(r_1, r_2)|p_1(t) > r_1 \text{ and } p_2(t) > r_2\},$$
$$F_1(t) = \{(r_1, r_2)|r_1 > p_1(t) \text{ and } r_1 - p_1(t) > r_2 - p_2(t)\},$$
$$F_2(t) = \{(r_1, r_2)|r_2 > p_2(t) \text{ and } r_2 - p_2(t) \geq r_1 - p_1(t)\}$$

represent customers' preferences on the duopolistic market at time t (see Figure 6.1).

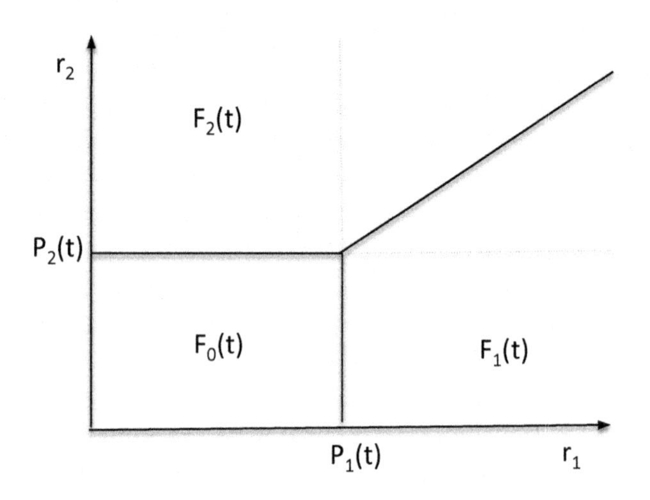

Fig. 6.1 Customer's preferences on the duopolistic market at time t

Suppose the number of customers $f_i(t)$ adopting service i at time t is proportional to the number of potential adopters $M - x_1(t) - x_2(t)$ and to the number of current subscribers of service i, $a_i + b_i\, x_i(t)$ with a_i and b_i representing advertising and word-of-mouth effects, respectively. Also, $f_i(t)$ is assumed to be proportional to the fraction of customers whose surplus from subscription to service i is higher than that from subscription to the other service, i.e.,

$$f_1(t) \sim W_1(p_1(t), p_2(t)) = \iint\limits_{r1, r2 \in F_1(t)} w(r_1, r_2)$$

$$= \int\limits_{p_1(t)}^{\infty} \int\limits_{0}^{r_1 p_2(t)/p_1(t)} w(r_1, r_2) dr_2 dr_1,$$

$$f_2(t) \sim W_2(p_1(t), p_2(t)) = \iint\limits_{r1, r2 \in F_2(t)} w(r_1, r_2)$$

$$= \int\limits_{p_2(t)}^{\infty} \int\limits_{0}^{r_2 p_1(t)/p_2(t)} w(r_1, r_2) dr_2 dr_1.$$

The number of customers who adopt service $i = 1, 2$, at time t is given by

$$\begin{aligned} f_i(t) &= f_i(x_1(t), x_2(t), p_1(t), p_2(t)) = \\ &= (M - x_1(t) - x_2(t))(a_i + b_i x_i(t)) W_i(p_1(t), p_2(t)). \end{aligned} \tag{6.1}$$

Under the assumption that $w(r_1, r_2) = w(r_1)w(r_2)$ (where $w(r_i)$, $i = 1, 2$, is the exponential probability density function with expectatiton $1/\alpha_i$), see [Maoui *et al.* (2009); Roth *et al.* (2006)], $W_1(p_1(t), p_2(t))$ and $W_2(p_1(t), p_2(t))$ become

$$W_1(p_1(t), p_2(t)) = \mathrm{e}^{-\alpha_1 p_1(t)} \left(1 - \frac{\alpha_1 p_1(t)\, \mathrm{e}^{-\alpha_2 p_2(t)}}{\alpha_1 p_1(t) + \alpha_2 p_2(t)}\right),$$

$$W_2(p_1(t), p_2(t)) = \mathrm{e}^{-\alpha_2 p_2(t)} \left(1 - \frac{\alpha_2 p_2(t)\, \mathrm{e}^{-\alpha_1 p_1(t)}}{\alpha_1 p_1(t) + \alpha_2 p_2(t)}\right).$$

Assume that a discontinuance rate $\gamma_i(t)$, $i = 1, 2$, for service i increases when subscription fee for service i increases compared to the competitor's subscription fee, i.e.:

$$\gamma_i(t) = \gamma_i(p_1(t), p_2(t)) = \gamma_i \frac{p_i(t)}{p_1(t) + p_2(t)},$$

where $\gamma_i > 0$. The diffusion of the demand for service i is modeled by

$$\dot{x}_i(t) = f_i(t) - \gamma_i(t)\, x_i(t). \tag{6.2}$$

The total profit from service i over the planning horizon $[0, T]$ is given by:

$$\Pi_i(x_1, x_2, p_i) = \int_0^T [x_i(t)(p_i(t) - c_i)]\, dt + S_i x_i(T), \quad i = 1, 2, \tag{6.3}$$

where $c_i \geq 0$ is a cost of subscription to service i and S_i is a salvage value of each customer at time T.

6.2 Nash Equilibrium

This section finds a Nash equilibrium for pricing strategies of both services such that no single competitor can benefit deviating from it. Nash equilibrium pricing strategies $\{p_1^*(t), p_2^*(t)\}$ are found from the Calculus of Variations problems [Giaquinta and Hildebrandt (1996); Gelfand and Fomin (1963)] with the variable endpoints for $x_i(t)$ at $t = T$:

$$\begin{aligned} \max_{p_1(t)} \Pi_1(x_1, x_2, p_1) \quad \text{s.t.} \quad & \dot{x_1}(t) = f_1(t) - \gamma_1(t)\, x_1(t), \quad x_1(0) = x_1^{(0)}, \\ t \in [0, T], \quad & \dot{x_2}(t) = f_2(t) - \gamma_2(t)\, x_2(t), \\ & x_2(0) = x_2^{(0)}, \quad t \in [0, T], \end{aligned} \tag{6.4}$$

$$\begin{aligned} \max_{p_2(t)} \Pi_2(x_1, x_2, p_2) \quad \text{s.t.} \quad & \dot{x_1}(t) = f_1(t) - \gamma_1(t)\, x_1(t), \quad x_1(0) = x_1^{(0)}, \\ t \in [0, T], \quad & \dot{x_2}(t) = f_2(t) - \gamma_2(t)\, x_2(t), \\ & x_2(0) = x_2^{(0)}, \quad t \in [0, T]. \end{aligned} \tag{6.5}$$

This problem can be solved by the Lagrange multipliers technique. The Lagrangians for (6.4) and (6.5) are given by

$$\begin{aligned} \mathcal{F}_1(x_1, x_2, p_1, \lambda_1(t), \mu_1(t)) &= \int_0^T \mathcal{L}_1(x_1(t), x_2(t), p_1(t), \lambda_1(t), \mu_1(t))dt \\ &\quad + S_1 x_1(T), \\ \mathcal{F}_2(x_1, x_2, , p_2, \lambda_2(t), \mu_2(t)) &= \int_0^T \mathcal{L}_2(x_1(t), x_2(t), p_2(t), \lambda_2(t), \mu_2(t))dt \\ &\quad + S_2 x_2(T), \end{aligned} \tag{6.6}$$

where

$$\begin{aligned} \mathcal{L}_1(x_1, x_1, p_1, \lambda_1, \mu_1) &= x_1(t)(p_1(t) - c_1) + \lambda_1(t)\left(\dot{x}_1(t) - f_1(t) + \gamma_1(t)\, x_1(t)\right) \\ &\quad + \mu_1(t)\left(\dot{x}_2(t) - f_2(t) + \gamma_2(t)\, x_2(t)\right), \\ \mathcal{L}_2(x_2, x_2, p_2, \lambda_2, \mu_2) &= x_2(t)(p_2(t) - c_2) + \lambda_2(t)\left(\dot{x}_2(t) - f_2(t) + \gamma_2(t)\, x_2(t)\right) \\ &\quad + \mu_2(t)\left(\dot{x}_1(t) - f_1(t) + \gamma_1(t)\, x_1(t)\right), \end{aligned}$$

and $\lambda_1(t)$, $\lambda_2(t)$, $\mu_1(t)$, $\mu_2(t)$ are Lagrange multipliers. The total variations of the functionals (6.6) are

$$\delta\mathcal{F}_1 = \int_0^T \left[\left(\frac{\partial \mathcal{L}_1}{\partial x_1} - \frac{d}{dt}\left(\frac{\partial \mathcal{L}_1}{\partial \dot{x}_1}\right)\right) \delta x_1 + \left(\frac{\partial \mathcal{L}_1}{\partial x_2} - \frac{d}{dt}\left(\frac{\partial \mathcal{L}_1}{\partial \dot{x}_2}\right)\right) \delta x_2 \right.$$
$$\left. + \frac{\partial \mathcal{L}_1}{\partial p_1}\delta p_1 + \frac{\partial \mathcal{L}_1}{\partial \lambda_1}\delta \lambda_1 + \frac{\partial \mathcal{L}_1}{\partial \mu_1}\delta \mu_1 \right] dt + [\lambda_1\, \delta x_1]|_0^T + S_1 x_1(T),$$

$$\delta\mathcal{F}_2 = \int_0^T \left[\left(\frac{\partial \mathcal{L}_2}{\partial x_1} - \frac{d}{dt}\left(\frac{\partial \mathcal{L}_2}{\partial \dot{x}_1}\right)\right) \delta x_1 + \left(\frac{\partial \mathcal{L}_2}{\partial x_2} - \frac{d}{dt}\left(\frac{\partial \mathcal{L}_2}{\partial \dot{x}_2}\right)\right) \delta x_2 \right.$$
$$\left. + \frac{\partial \mathcal{L}_2}{\partial p_2}\delta p_2 + \frac{\partial \mathcal{L}_2}{\partial \lambda_2}\delta \lambda_2 + \frac{\partial \mathcal{L}_2}{\partial \mu_2}\delta \mu_2 \right] dt + [\lambda_2\, \delta x_2]|_0^T + S_2 x_2(T).$$

The necessary conditions for existence of the Nash equilibrium $(p_1^*(t), p_2^*(t))$ are given by

$$\delta\mathcal{F}_1 = 0, \quad \delta\mathcal{F}_2 = 0,$$

that reduce to a system of differential equations with the initial conditions

$$x_1(0) = x_1^{(0)}, \quad x_2(0) = x_2^{(0)}, \tag{6.7}$$

and the transversality conditions

$$\begin{aligned} \lambda_1(T) &= -S_1, \quad \mu_1(T) = 0, \\ \lambda_2(T) &= -S_2, \quad \mu_2(T) = 0. \end{aligned} \tag{6.8}$$

The next section illustrates the model and Nash equilibrium pricing strategies numerically.

6.3 Numerical Examples

6.3.1 *Example 1*

This example shows how adoption of services 1 and 2 at time $t = t_0$ depend on subscription fees $p_1 = p_1(t_0)$ and $p_2 = p_2(t_0)$. Let an average fee that customers are willing to pay for a service subscription be \$25 (i.e. $\alpha_1 = \alpha_2 = 0.04$). Figures 6.2 (a) and 6.2 (b) show that $W_1 = W_1(p_1(t_0), p_2(t_0))$ and $W_2 = W_2(p_1(t_0), p_2(t_0))$ are decreasing functions of $p_1(t_0)$ and $p_2(t_0)$ and are increasing functions of $p_2(t_0)$ and $p_1(t_0)$, respectively.

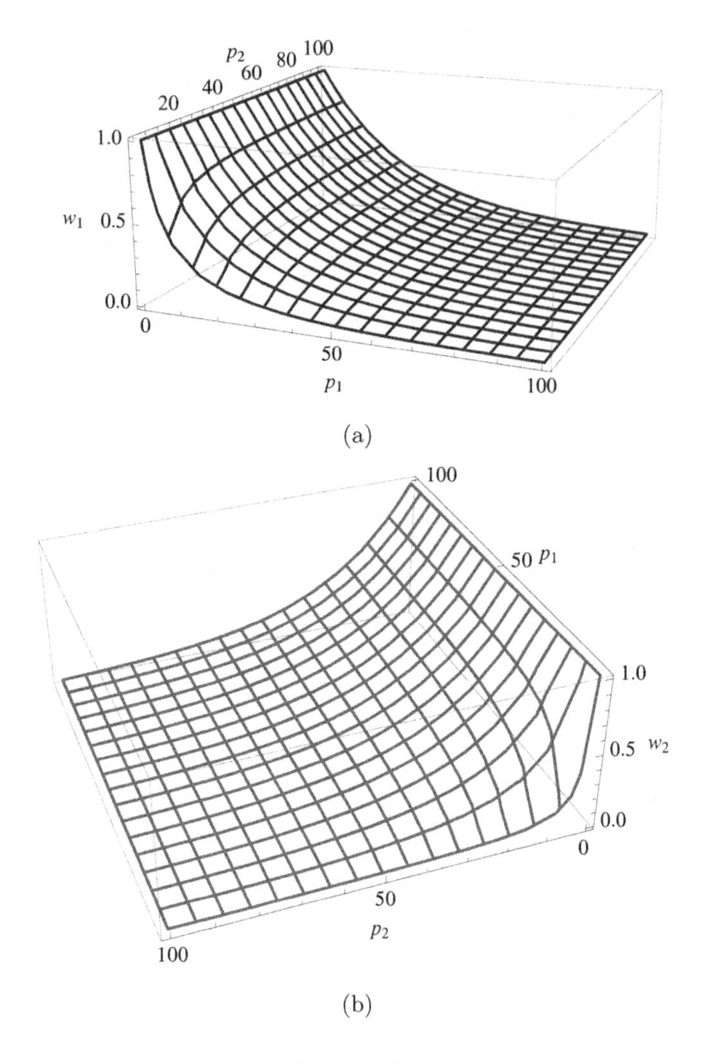

Fig. 6.2 Shares of the potential market subscribing to the services vs. subscription fees

6.3.2 *Example 2*

This example illustrates the model (6.2) when both pricing strategies $p_1(t)$ and $p_2(t)$ are known. Let the market have 100 potential adopters ($M = 100$) with 5 customers subscribed to each service at $t = 0$ ($x_1^{(0)} = x_2^{(0)} = 5$) with $T = 1$. Parameters associated with advertising and word-of-mouth effects, costs of subscription, churn rate parameters, and reservation prices are assumed to be the same for both services: $a_1 = a_2 = b_1 = b_2 = 0.5$,

Table 6.1 Values of parameters in Example 2

M	$x_1^{(0)}$	$x_2^{(0)}$	c_1	c_2	a_1	a_2	b_1	b_2	α_1	α_2	γ_1	γ_2
100	5	5	10	10	0.5	0.5	0.5	0.5	0.04	0.04	0.2	0.2

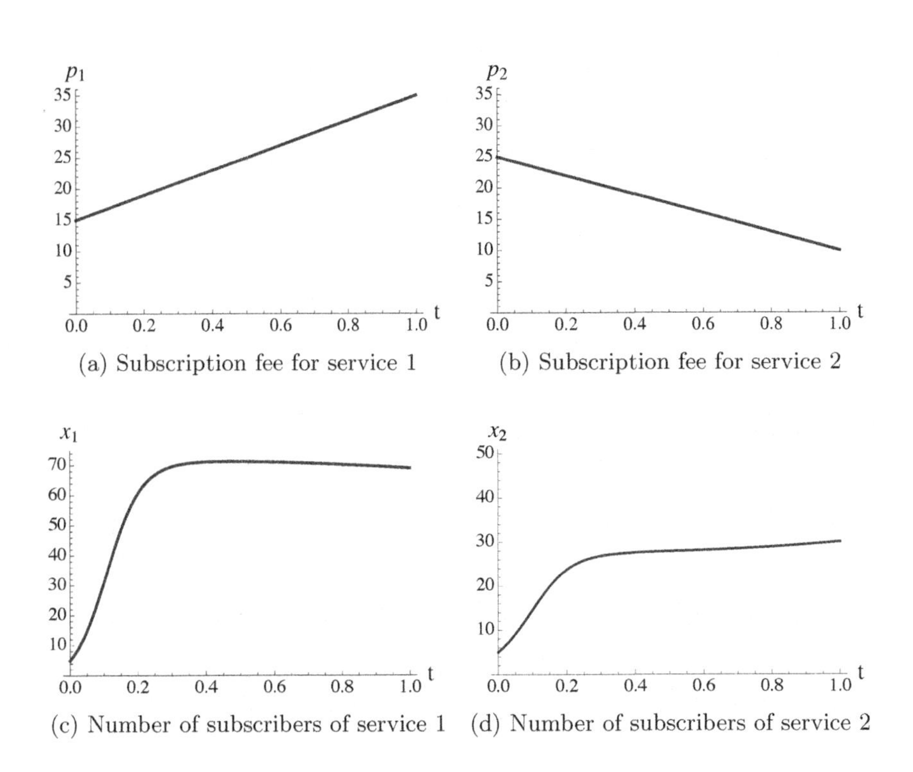

(a) Subscription fee for service 1

(b) Subscription fee for service 2

(c) Number of subscribers of service 1

(d) Number of subscribers of service 2

Fig. 6.3 Forecasted number of subscribers to services 1 and 2

$c_1 = c_2 = 10$, $\gamma_1 = \gamma_2 = 0.2$, and $\alpha_1 = \alpha_2$, see Table 6.1. It is assumed that service providers have different pricing strategies: the provider of service 1 has a penetration pricing strategy ($p(t) = 15 + 20t$, see Figure 6.3 (a)) whereas the provider of service 2 has a skimming pricing strategy ($p(t) = 25 - 15t$, see Figure 6.3 (b)). Figures 6.3 (c)–(d) show the the total numbers of customers subscribing to services 1 and 2 over the period $[0, 1]$. In the beginning of the planning horizon, majority of customers adopted service 1, because service 1 was cheaper. However, later when service 2 became cheaper, subscribers of service 1 migrated to service 2.

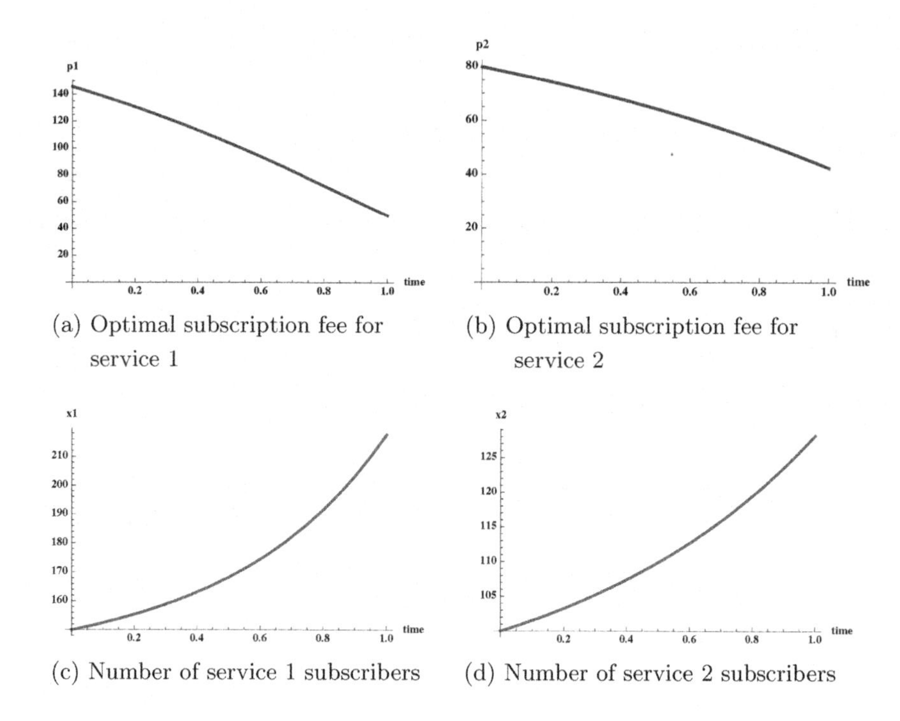

(a) Optimal subscription fee for service 1

(b) Optimal subscription fee for service 2

(c) Number of service 1 subscribers

(d) Number of service 2 subscribers

Fig. 6.4 Optimal subscription fee strategies in duopoly

6.3.3 *Example 3*

This example illustrates a Nash equilibrium for pricing strategies developed in Section 6.2. Suppose there are 1000 customers ($M = 1000$) on the market with 150 and 100 initial subscribers of services 1 and 2, respectively ($x_1^{(0)} = 150$ and $x_2^{(0)} = 100$), and suppose advertising and word-of-mouth effects as well as customers' reservation prices for service 1 are higher than those for service 2. Table 6.2 presents parameter values.

Figures 6.4 (a) and 6.4 (c) show the optimal subscription fee and total numbers of subscribers as functions of time for service 1, whereas Figures 6.4 (b) and 6.4 (d) illustrate the same functions for service 2. This example shows that skimming pricing strategies are optimal for both providers. At the introductory stage, the optimal subscription fee for service 1 is higher than that for service 2, and in the end of the planning horizon, the subscription fees for the both services are approximately the same.

Table 6.2 Values of parameters in Example 3

M	$x_1^{(0)}$	$x_2^{(0)}$	S_1	S_2	c_1	c_2
1000	150	100	40	40	10	10

a_1	a_2	b_1	b_2	α_1	α_2	γ_1	γ_2
0.6	0.5	0.6	0.5	0.02	0.04	0.05	0.05

6.3.4 *Example 4*

This example applies the model (6.2) to the satellite radio industry, which was a duopoly prior to a merger of former competitors Sirius and XM in 2008. Sirius and XM charged a monthly fee for delivering multiple entertainment channels with a limited number commercials if any at all. Satellite radio is especially beneficial for long-distance drivers, since satellite signal does not fade out no matter how far they travel within the US. Car manufacturers install satellite-ready radios as a standard feature of multiple car models.

This example uses yearly data for the average revenue per subscriber and for the total number of subscribers for Sirius and XM satellite radio providers during 2003-2007 years (see Figure 6.5). The data was collected from press releases [Sirius Satellite Radio (2010); XM Satellite Radio (2010)] available online. The average monthly revenue per subscriber (ARPU) is calculated as the total earned subscription revenue and activation revenue during the period, over the daily weighted average number of subscribers for the period.

In this example, interpolation curves of ARPU for Sirius and XM (see Figure 6.6) are used as substitutes of monthly subscription fees $p_1(t)$ and $p_2(t)$ in the model (6.2). For illustration assume the following: (i) a potential market consists of 20 million customers, (ii) on average customers are willing to pay more for subscription to Sirius radio than to XM radio, and (iii) advertising and word-of mouth effects for Sirius radio are higher than those for XM radio (see values of parameters given by Table 6.3). Figure 6.6 (b) shows the total number of subscribers for both companies obtained from the model (6.2), where dots represent real data. This shows that the model (6.2) for services diffusion agrees with real life data reasonably well.

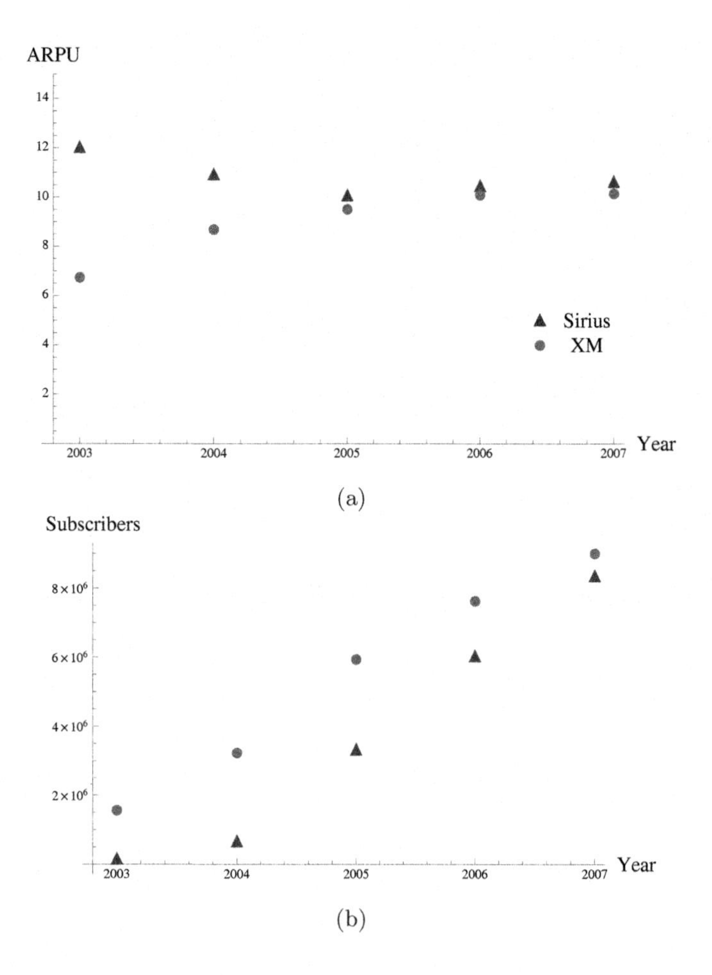

Fig. 6.5 (a) Average monthly revenue per customer (ARPU) for Sirius and XM; (b) Total number of Sirius and XM subscribers

Table 6.3 Values of parameters in Example 4

M	$x_1^{(0)}$	$x_2^{(0)}$	a_1	a_2
$2 * 10^7$	149612	1560228	0.2	0.18

b_1	b_2	α_1	α_2	γ_1	γ_2
0.01	0.08	0.05	0.08	0.04	0.04

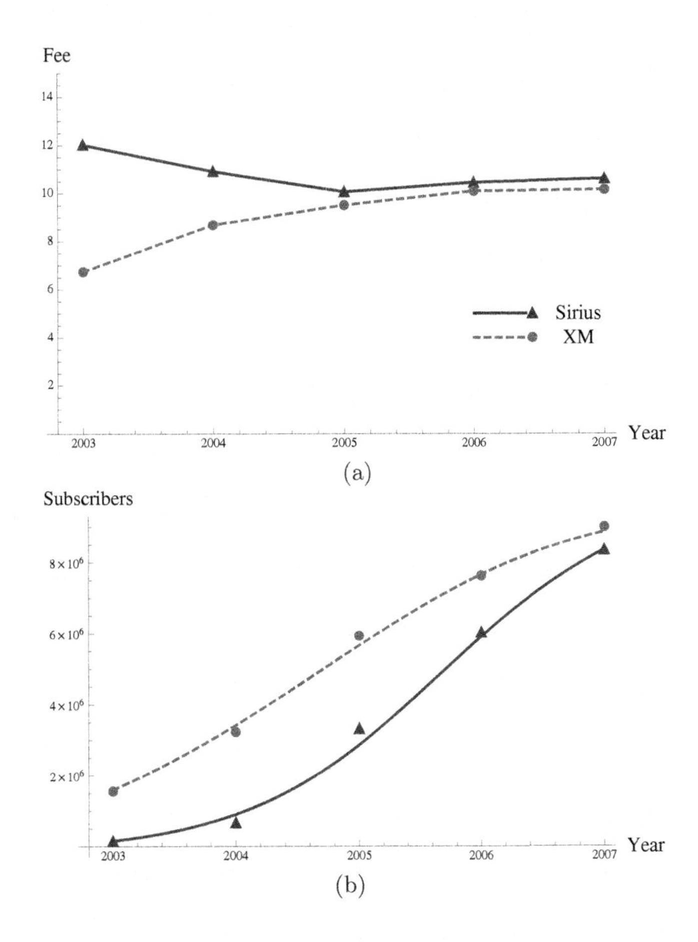

Fig. 6.6 (a) Substitutes of monthly subscription fees for Sirius and XM; (b) Total number of Sirius and XM subscribers forecasted by the model

6.4 Conclusions

The chapter has developed the model for dynamic optimal pricing of two services on the duopolistic market. The model considers advertising and word-of-mouth effects along with customer churn and assumes the following: (i) subscription fee is continuous; (ii) reservation fees for both providers are independently distributed; (iii) distributions for reservation fees are independent of the number of potential adopters. It shows that the optimal subscription fee strategies for two competative services has a Nash equilibrium. In addition, numerical examples suggest skimming subscription fee strategies for both providers.

Chapter 7

Optimal Investment and Pricing Strategies for Post-Production Service Contracts

7.1 Overview

In the defense industry, equipment sustainment expenses often exceed costs of the equipment itself. For example, an expected cost to sustain the Joint Strike Fighter exceeds its development and production cost by over $250 billion [Government Accountability Office (2008)]. As an efficient product support strategy reducing maintenance expenses, performance-based contracting (PBC) has become a topic of growing interest especially in the defense industry. In 2001, the United States Department of Defense (DoD) stated that PBC would be their preferred method for procuring maintenance support [Vitasek and Geary (2008)]. Currently, the DoD is engaged in 76 performance-based contracts with another 95 scheduled in the near future [Geary and Vitasek (2008)]. PBC has also been successfully employed in the commercial sector including aerospace, transportation, telecommunication, and power generation industries [Keating and Huff (2005)]. By 2005, 50 countries were exploring or implementing PBC [Program (2009)]. Existing practices indicate that PBC reduces maintenance costs and improves system performance [Kratz (2008); Program (2009)].

Suppliers that transition from the traditional post-production service contracting to PBC face a number of challenging management decisions: investment strategy, targeted level of reliability, contract's length and pricing, to name just a few. Undoubtedly, system quality (reliability) is highly correlated with investment: on the one hand, it reduces maintenance costs, but on the other hand, it leads to high service contract fees. Although customers value improved system reliability, they may not engage in PBC due to its high fees. Thus, a supplier maximizes its profit by trading off investment, system reliability, as well as the contract's length and fees.

Despite the growing attention to PBC in the existing literature, most research publications in this area are qualitative and have limited focus on business aspects of PBC. They describe current practices in PBC [Keating and Huff (2005)]; discuss advantages of PBC over traditional contracting [Kim *et al.* (2007)]; address impact of PBC on system reliability [Kim *et al.* (2009)]; formulate more efficient and effective PBC agreements [Sols *et al.* (2007)] and reward schemes in PBC [Sols *et al.* (2008)]; and develop inventory allocation models [Nowicki *et al.* (2008)]. However, none of the existing works develop optimal investment and pricing strategies for PBC. This chapter bridges the gap. It develops a decision-theoretic model for the optimal investment strategy, pricing strategy and length of the post-production service contract that maximize profit of the supplier offering PBC. The model assumes that customers' willingness to pay for the contract depends on contract's length and system reliability.

The chapter is organized as follows. Section 7.2 develops a decision-theoretic model for a supplier offering PBC. Section 7.3 derives supplier's optimal investment and pricing strategies. Section 7.4 illustrates optimal strategies numerically, and Section 7.5 concludes the chapter.

7.2 Model

Suppose a supplier sells a system and offers customers an option of a k-period post-production maintenance service contract at a fixed periodical fee p. Under the contract, the supplier is responsible for all costs and risks associated with sustaining the proper operation of the system. The system has an initial reliability r_0, and the supplier has the ability to improve the system's design by investing capital x in system reliability $r(x) \geq r_0$. Suppose a market consists of M potential customers whose willingness to pay the periodic fee depends on $r(x)$ and k. Let $w_{r,k}(v)$, $v > 0$, be a probability density function of reservation fees (maximal fees that customers are willing to pay periodically) for the k-period contract if the system reliability is r. A customer buys the post-production service contract if the actual periodic contract fee p is less than or equal to the customer's reservation fee. The fraction of the potential market that will engage in the k-period post-production service contract is determined by

$$W_{r,k}(p) = \int_{p}^{\infty} w_{r,k}(v)dv. \tag{7.1}$$

With capital x invested in system reliability, the total profit of the supplier is given by:

$$\Pi(x,p,k) = M\sum_{j=1}^{k}\frac{1}{(1+i)^j}(p - f(r(x)))\int_{p}^{\infty} w_{r(x),k}(v)dv - x, \tag{7.2}$$

where i is an interest rate, and $f(r(x))$ is an expected cost of all system failures for a single period.

7.2.1 *Model Assumptions*

The model makes the following assumptions:

(A1) The function $r(x)$ satisfies the initial reliability condition $r(0) = r_0$ and has a signoid shape (as observed in reality [Levesque (2000)]):

$$r(x) = r_0 + (1 - r_0)\left(1 - \frac{1}{x/\gamma + 1}\right) = \frac{x + r_0\gamma}{x + \gamma}, \tag{7.3}$$

where $\gamma > 0$ is a marginal investment parameter.

(A2) The cost per failure is a normally distributed random variable with the mean μ_c and variance σ_c^2.

(A3) The expected cost of all system failures per period decreases with reliability improvements, i.e.

$$f(r) = \mu_c m(1 - r),$$

where m is the number of missions in a single time period.

(A4) Customers' reservation fees follow the triangular distribution:

$$w_{r,k}(v) = \begin{cases} \frac{(\lambda(1-d(k-1))r-p)^2}{(\lambda(1-d(k-1))r)^2}, & 0 \leq p \leq \lambda(1 - d(k-1))r, \\ 0, & \text{otherwise}, \end{cases}$$

where λ is a maximal fee that customers are willing to pay for a single-period contract with $r(x) = 1$, and d is a discount per period expected by customers buying a multi-period contract.

7.3 Optimization

The supplier maximizes the expected profit $E[\Pi(x,p,k)]$ from a k-period contract ($k = 1,\ldots,n$) with respect to investment x^*, periodic contract fee

p^*, and contract length k^*:

$$E[\Pi(x^*, p^*, k^*)] = \max_{k=1,\dots,n} E[\Pi(x^*, p^*, k)], \tag{7.4}$$

where

$$E[\Pi(x^*, p^*, k)] = \max_{\{x,p\}\in F_{xp}} E[\Pi(x, p, k)] \tag{7.5}$$

with a set of feasible solutions:

$$F_{xp} = \{\{x, p\}|x \geq 0,\ 0 \leq p \leq \lambda(1 - d(k-1))r\}.$$

Under the assumptions (A1)–(A4), the expected profit takes the form

Table 7.1 Baseline Example

Parameter	d	M	λ	r_0	γ	μ_c	σ_c	m	i
Value	0.07	60	100	0.7	100	20	10	10	0.05

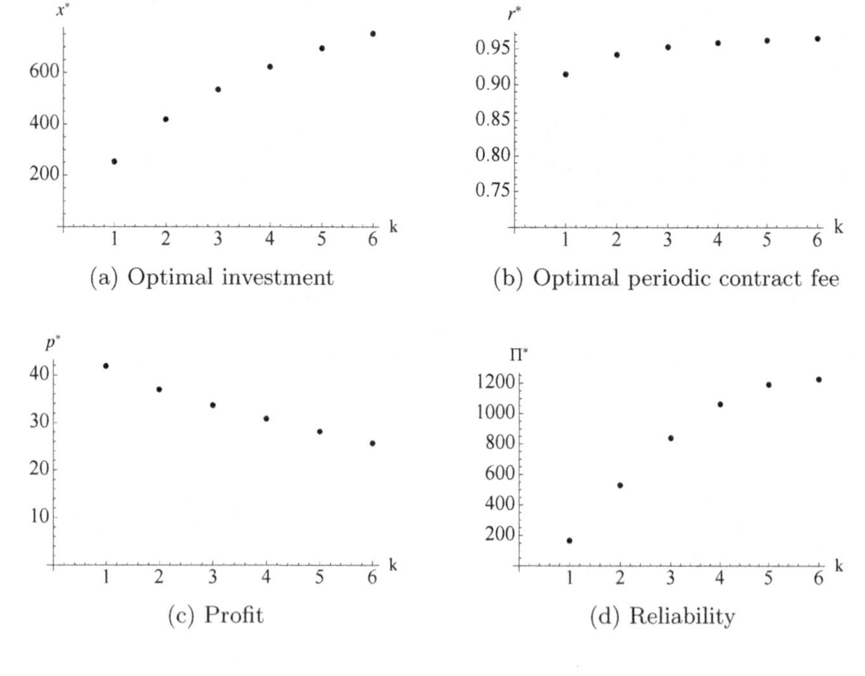

Fig. 7.1 Optimal investment, reliability, periodic contract fee and profit as functions of the length of a contract

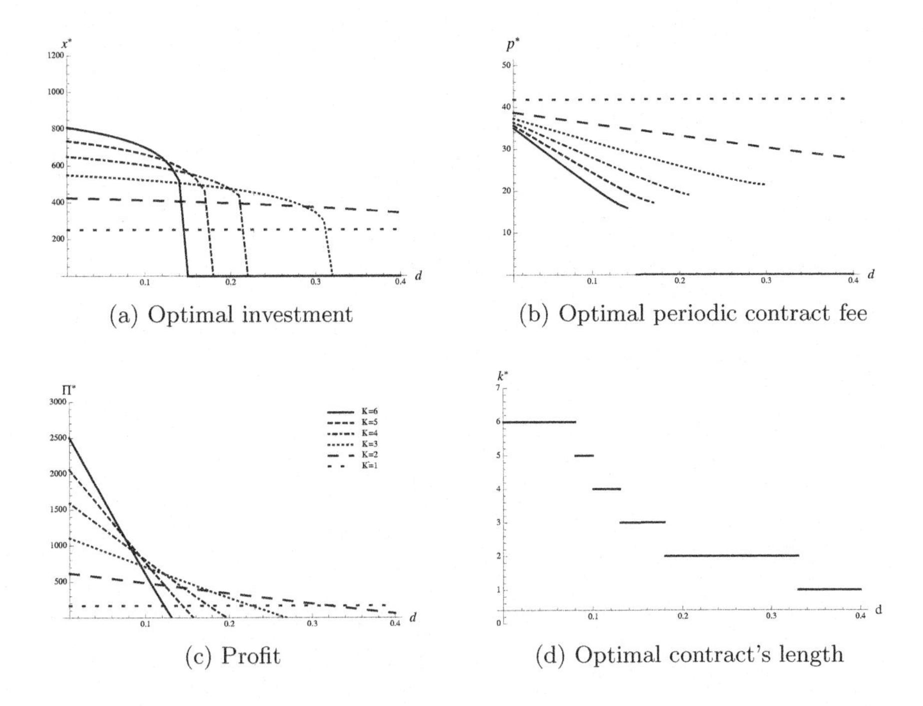

(a) Optimal investment

(b) Optimal periodic contract fee

(c) Profit

(d) Optimal contract's length

Fig. 7.2 Optimal investment, periodic contract fee, profit and optimal contract's length as functions of the discount expected by customers

$$E[\Pi(x,p,k)] = \begin{cases} \dfrac{M\, I_k\, (p(x+\gamma) - \mu_c m\,(1-r_0)\,\gamma)\,(p(x+\gamma) - \lambda D_k(x+r_0\gamma))^2}{\lambda^2 D_k^2 (x+r_0\gamma)^2 (x+\gamma)} - x, & \\ & 0 \le p \le \lambda D_k\, r(x), \\ 0, & \text{otherwise}, \end{cases} \tag{7.6}$$

where $D_k = (1 - d(k-1))$ and $I_k = (1 + i - (1+i)^{-k})/i$.

The optimal investment x^* and the optimal periodic fee p^* for the k-period contract are either critical points determined from the first order necessary conditions:

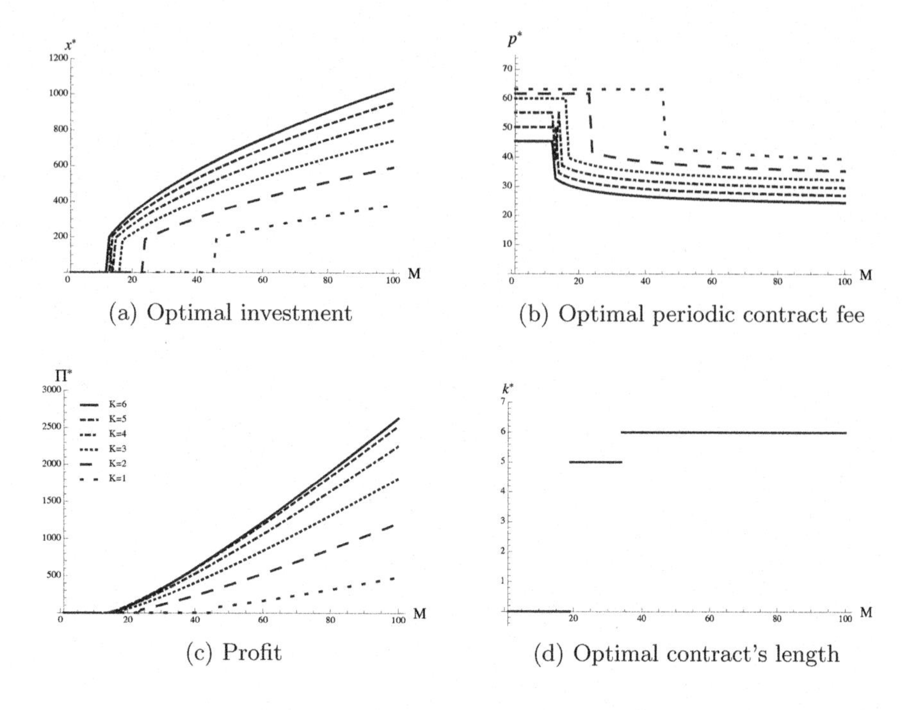

(a) Optimal investment

(b) Optimal periodic contract fee

(c) Profit

(d) Optimal contract's length

Fig. 7.3 Optimal investment, periodic contract fee, profit and optimal contract's length as functions of the market size

$$
\begin{aligned}
\left.\frac{\partial E[\Pi(x,p,k)]}{\partial x}\right|_{(x^*,p^*,k)} &= 0, \\
\left.\frac{\partial E[\Pi(x,p,k)]}{\partial p}\right|_{(x^*,p^*,k)} &= 0,
\end{aligned}
\tag{7.7}
$$

or belong to the boundary of the feasible set F_{xp}. With (7.6), (7.7) reduces to

$$
\begin{aligned}
& p = \frac{2\,\mu_c\, m\,(1-r_0)\,\gamma + \lambda\, D_k\, X}{3\,(X - \gamma\,(1-r_0))}, \\
& 4M\,\gamma\, I_k\,(1-r_0)\,(X\,\lambda\, D_k - \mu_c\, m(1-r_0)\gamma)^{\,2}\,(\mu_c\, m(3X + 2(1-r_0)\gamma) \\
& \quad + X\,\lambda\, D_k) - 27X^3\lambda^2 D_k^2 (X + (1-r_0)\gamma)^2 = 0,
\end{aligned}
\tag{7.8}
$$

where $X = x + r_0\gamma$. If (x^*, p^*) is a critical point, it should satisfy the second order sufficient conditions:

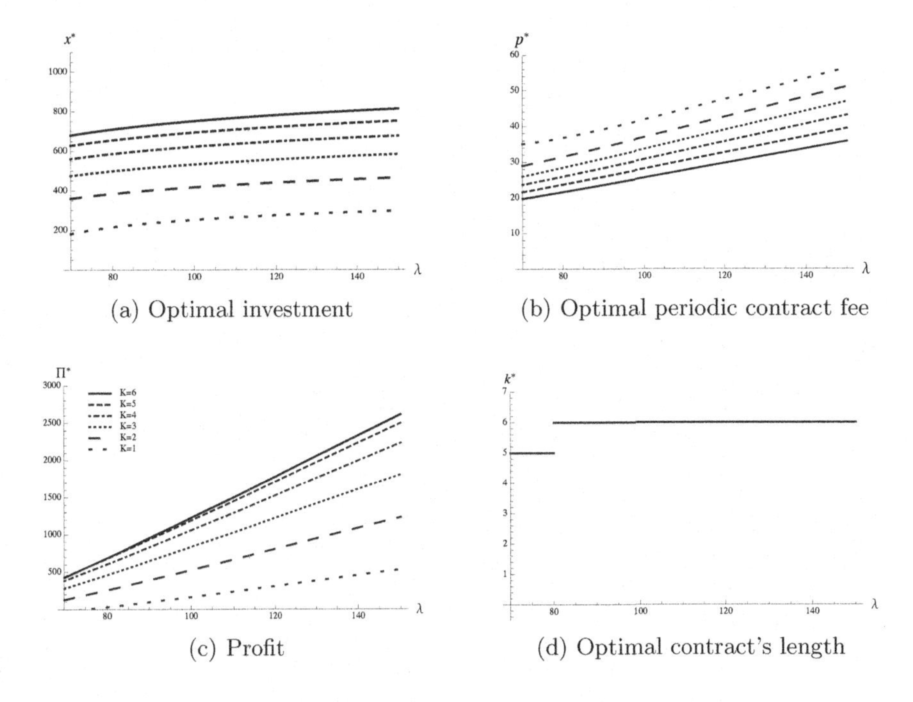

(a) Optimal investment

(b) Optimal periodic contract fee

(c) Profit

(d) Optimal contract's length

Fig. 7.4 Optimal investment, periodic contract fee, profit and optimal contract's length as functions of the maximal price that customers are willing to pay for a single-period contract

$$\left.\frac{\partial^2 E[\Pi(x,p)]}{\partial^2 x}\right|_{(x^*,p^*)} < 0,$$
$$\left.\frac{\partial^2 E[\Pi(x,p)]}{\partial^2 p}\right|_{(x^*,p^*)} < 0, \tag{7.9}$$

$$\left.\frac{\partial^2 \Pi(x,p)}{\partial^2 x}\frac{\partial^2 \Pi(x,p)}{\partial^2 p} - \frac{\partial^2 \Pi(x,p)}{\partial x \partial p}\frac{\partial^2 \Pi(x,p)}{\partial p \partial x}\right|_{(x^*,p^*)} > 0. \tag{7.10}$$

The solution (x^*, p^*) is obtained numerically for all $k = 1, \ldots, n$, and the optimal contracting period k^* follows from (7.4).

7.4 Numerical Example

This section analyzes the optimal investment x^*, optimal contract fee p^*, reliability $r(x^*)$, and the expected profit $\Pi^* = E[\Pi(x^*, p^*, k)]$ as functions of the contract length k and parameters d, λ, μ_c, r_0, M, and γ.

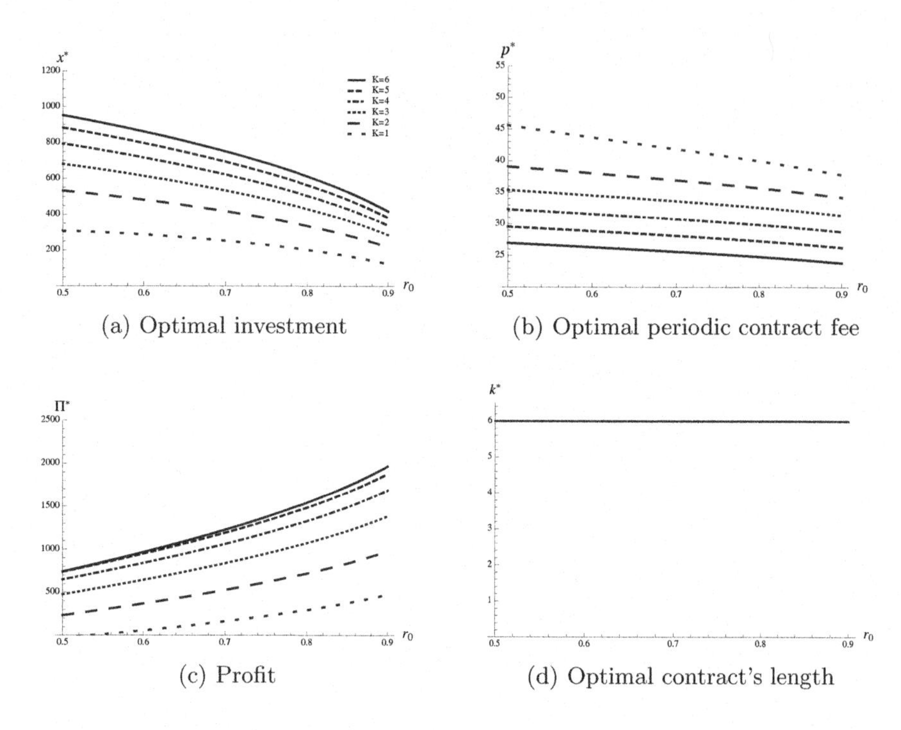

Fig. 7.5 Optimal investment, periodic contract fee, profit and optimal contract's length as functions of the initial reliability

Suppose a supplier of airplane engines offers a performance-based post-production service option to 60 potential customers ($M = 60$). The maximal fee that customers are willing to pay periodically for the post-production maintenance service contract (if reliability of the engines is improved to 1) is \$100, 000. Customers expect to have 7% discount per period ($d = 0.07$) if they subscribe to a multi-period contract. The initial reliability of the engines is 0.7 ($r_0 = 0.7$) and at least \$100, 000 of investment is required to improve reliability of the engines up to $r_0 + \frac{1}{2}(1 - r_0) = 0.85$ ($\gamma = 100$). Let the periodic interest rate be 5% ($i = 0.05$). The expected cost and variance per failure are \$20, 000 ($\mu_c = 20$) and \$10, 000 ($\sigma_c^2 = 10$), respectively. There are 10 missions per period ($m = 10$). Table 7.1 summarizes parameters in this example.

The optimal investment and optimal contract fee for the k-period contract ($k = 1, \ldots, 6$) are determined by (7.8), see Figures 7.1 (a) and (b). Now suppose the parameters in Table 7.1 vary. Figures 7.2–7.7 show x^*, p^*, Π^*, and k^* as functions of the discount per period, d; market size, M; customers' willingness to pay, λ; initial reliability, r_0; marginal investment

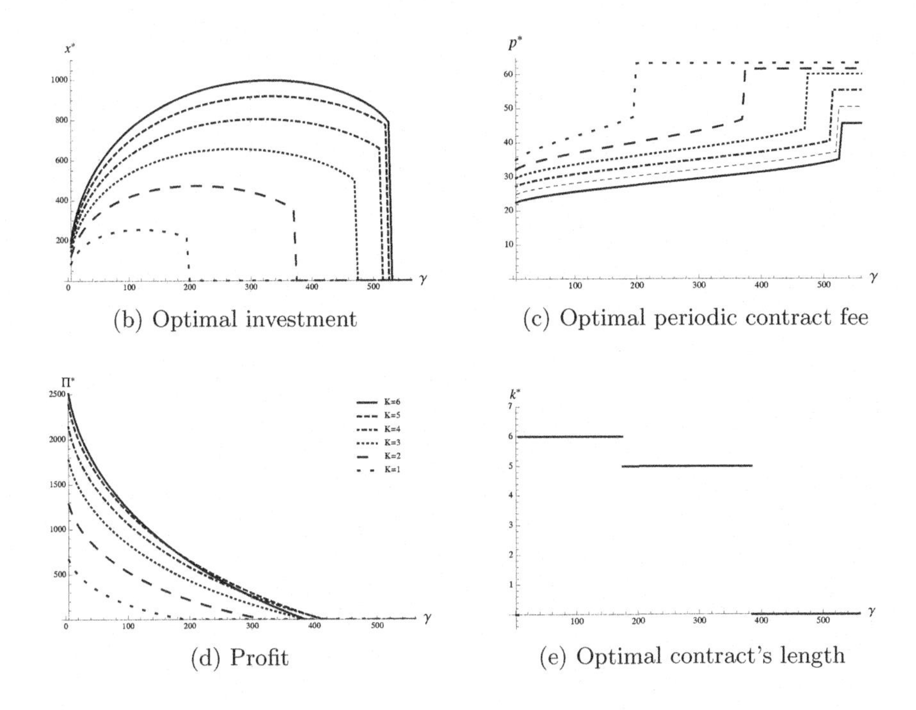

(b) Optimal investment

(c) Optimal periodic contract fee

(d) Profit

(e) Optimal contract's length

Fig. 7.6 Optimal investment, periodic contract fee, profit and optimal contract's length as functions of the marginal investment parameter

parameter, γ; and the expected cost per failure, μ_c. Thus, the optimal contract's length $k^* = 6$ if $d \in [0, 0.08]$, $M \in [38, 100]$, $\lambda \in [80, 150]$, $r_0 \in [0.5, 0.9]$, $\gamma \in [0, 170)$ and $\mu_c \in [0, 34)$; $k^* = 5$ if $d \in (0.08, 0.1]$, $M \in [20, 38)$, $\lambda \in [60, 80]$, $\gamma \in [170, 380)$, $\mu_c \in [34, 40)$; $k^* = 4$ if $d \in (0.1, 0.13]$, $k^* = 3$ if $d \in (0.13, 0.18]$, $k^* = 2$ if $d \in (0.18, 0.33]$, $k^* = 1$ if $d \in [0.33, 0.4]$; and finally PBC is unprofitable if $\gamma \in [380, +\infty)$ and $M \in [0, 20)$.

In summary, the following conclusions can be drawn:

- Optimal investment is an increasing function of the expected cost per failure, market size, and customers' willingness to pay, but it is a decreasing function of the initial reliability.
- Optimal periodic contract fee is an increasing function of the contract' length, customers' willingness to pay, and an expected cost per failure, but it is a decreasing function of the initial reliability and market size.
- Longer post-production service contracts require higher optimal investments but provide higher system reliability.

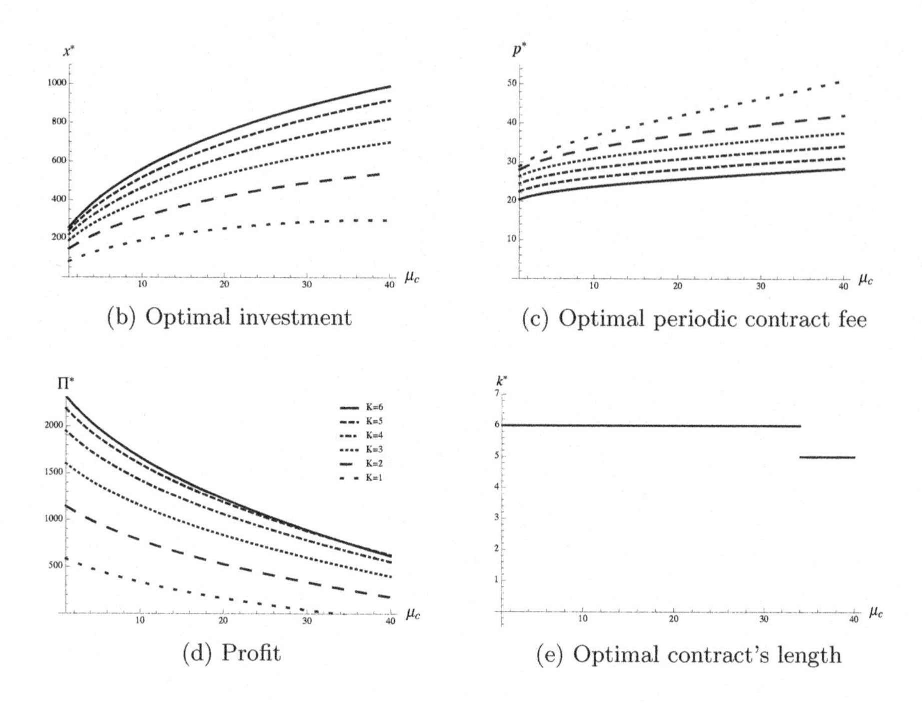

Fig. 7.7 Optimal investment, periodic contract fee, profit and optimal contract's length as functions of the expected cost per failure

- Optimal contract length is a decreasing function of the discount per period.

7.5 Conclusions

This chapter has developed a model for the optimal investment and pricing strategies for performance-based post-production service contracts. The model maximizes profit of a supplier with respect to contract's length, investment, and periodic contract fee. It assumes that customers' willingness to pay depends on contract's length and system reliability, and that the supplier can improve system reliability. Numerical examples have analyzed optimal contract's length, investment, system reliability, and optimal periodic fee with respect to the initial system reliability, customers' willingness to pay, the expected cost per failure, and with respect to other parameters of the model.

Chapter 8

Conclusions

New product development process requires a number of decisions to be made well in advance of the product's launch and throughout the entire product's life-cycle. Those decisions include: investment, scheduling, targeted reliability, infrastructure size, etc. Their success depends on accuracy of projections for future demand and on new product diffusion models. Constant advances in the telecommunication industry stimulate an extensive growth of new subscription services and pose several research questions. How does demand for a new service evolve in time? What does affect demand for a new service? How does demand for a new service affect demand for existing services? What is the relationship between individual adoption decisions and aggregate service diffusion?

The book has addressed these questions and has analysed demand for a new service on macro (company) and micro (individual customer) levels. The macro level analysis develops analytical models for diffusion of a new subscription services on monopolistic and duopolistic markets. The models account for advertising and word-of-mouth effects, customers' willingness to pay for services and also for customers' churn. Parameters of the models can either be estimated from early data or be dimensioned from comparable services introduced in the past. The models provide an insight on the role of marketing variables and competition in diffusion of a new service and can help managers to better control the diffusion and to optimize new service investment and pricing strategies. The micro level analysis deals with customers' migration from a legacy service to a technologically innovative substitute, forecasts future migration behavior of each customer, and segments customers based on their migration behavior. The analysis guides managers in making decisions on adoption of new technologies, infrastructure planning, and on targeting customers for marketing campaigns.

Another major contribution of the book is the first quantitative model for performance-based post-production service contracts resulting in the optimal investment and pricing strategies. The developed model is invaluable to suppliers in evaluating the economic viability of transitioning from traditional to a performance-based service contracting. It also shows how optimal strategies, system reliability, and profit depend on market size, customers' willingness to pay, expected cost per failure, marginal investment parameter, and contract length.

As a broader impact, the book deepens understanding of new service diffusion and customer migration from a legacy services to a new substitute. The developed models provide accurate projections of service demand, ensure efficient targeted marketing campaigns and increase companies' profits by trading off investment, reliability, and price of a new service.

Bibliography

Allenet, B. and Barry, H. (2003). Opinion and behaviour of pharmacists towards the substitution of branded drugs by generic drugs: Survey of 1,000 French community pharmacists, *Pharmacy World and Science* **25**, 5, pp. 197–202.

Arrow, K. J. and Kurz, M. (1970). *Public Investment, the Rate of Return, and Optimal Fiscal Policy* (John Hopkins, Baltimore).

Bass, F. (1969). A new product growth model for consumer durables, *Management Science*, 15, pp. 215–227.

Birge, J., Drogos, J. and Duenya, I. (1998). Setting single-period optimal capacity levels and prices for substitutable products, *The International Journal of Flexible Manufacturing Systems* **10**.

Breidert, C. (2006). *Estimation of Willingness-to-Pay Theory, Measurement, Application*, Chap. Willingness-to-Pay (WTP) in Marketing (Willingness-to-Pay (WTP) in Marketing).

California State University Northridge online (2009). Television statistics, URL `http://www.csun.edu/science/health/docs/tv&health.html`.

Chintagunta, P. K. and Rao, V. R. (1996). Pricing strategies in a dynamic duopoly: A differential game model, *Management Science* **42**, 11, pp. 1501–1514.

Chopra, V. (2005). The internet and mail, A Pitney Bowes Background paper for the project "Electronic Substitution for Mail: Models and Results, Myth and Reality", URL `http://www.postinsight.com`.

Cohen, M. A., Eliashberg, J. and Ho, T.-H. (1996). New product development: The performance and time-to-market tradeoff, *Management Science* **42**, 2, p. 173.

Constantiou, I. D. and Kautz, K. (2008). Economic factors and diffusion of IP telephony: Empirical evidence from an advanced market, *Telecommunications Policy* **32**, pp. 197–211.

Defense Acquisition University (2005a). Defense acquisition guidebook, URL http://akss.dau.mil/dag.

Defense Acquisition University (2005b). Performance based logistics: A program manager's product support guide.

Dekimpe, M. G., Parker, P. M. and Sarvary, M. (1998). Staged estimation of international diffusion models: An application to global cellular telephone adoption, *Technological Forecasting and Social Change* **57**, 1–2, pp. 105–132.

Deshmukh, S. D. and Chikte, S. D. (1977). Dynamic investment strategies for a risky R and D project, *Journal of Applied Probability* **14**, 1, pp. 144–152.

D'Este, G. M. (1981). A mMorgenstern-type bivariate gamma distribution, *Biometrica* **68**, 1, pp. 339–340.

Dockner, E. (1985). Optimal pricing in dynamic duopoly game model, *Zeitschrift fur Operations Research Series B* **29**, pp. 1–16.

Dockner, E. and Jorgensen, S. (1988). Optimal pricing strategies for new products in dynamic oligopolies, *Marketing Science* **7**, 4, pp. 315–334.

Dodds, W. (1973). An application of the Bass model in long-term new product forecasting, *Journal of Marketing Research*, 10, pp. 308–311.

Dolan, R. and Jeuland, A. (1981). Experience curves and dynamic demand models: Implications for optimal pricing strategies, *Journal of Marketing*, 45, pp. 52–62.

E. Asgharizadeh, D. M. (2000). Service contracts: A stochastic model, *Mathematical and Computer Modelling* **31**, 10–12.

Edwards, J. (2008). Drug marketing poised for historic decline, URL `http://www.brandweek.com`.

Efron, B. (1979). Bootstrap methods: Another look at the jackknife, *Annals of Statistics* **7**.

Feichtinger, G. and Dockner, E. (1985). Optimal pricing in duopoly: A noncooperative differential games solution, *Journal of Optimization Theory and Applications* **42**, 2, pp. 199–218.

Feng, Y. and Gallego, G. (2000). Perishable asset revenue management with Markovian time dependent demand intensities, *Management Science* **46**, 7, pp. 941–956.

Fisher, J. C. and Pry, R. H. (1971). A simple substitution model of technological change, *Technological Forecasting and Social Changes* **3**, pp. 75–88.

For Office of Financial Management, F. G. (2005). Best practices and trends in performance based contracting, .

Fruchter, G. E. and Rao, R. C. (2001). Optimal membership fee and usage price over time for a network service, *Journal of Service Research* **4**, 1, pp. 3–14.

Geary, S. and Vitasek, K. (2008). *Performance-Based Logistics: A Contractor's Guide to Life Cycle Product Support Management* (Bellevue, WA: Supply Chain Visions).

Geering, H. P. (2007). *Optimal Control with Engineering Applications* (Springer, Berlin Heldelberg).

Gelfand, I. M. and Fomin, S. V. (1963). *Calculus of Variations* (Prentice Hall, Englewood Cliffs, New Jersey).

Giaquinta, M. and Hildebrandt, S. (1996). *Calculus of Variations I. The Lagrangian Formalism* (Springer-Verlag).

Government Accountability Office (2008). Joint strike fighter recent decisions by department of defense add to program risks, In G. A. Office (Ed.), Vol. GAO-08-388. Washington D.C.

Granot, D., Granot, F. and Mantin, B. (2007). A dynamic pricing model under duopoly competition, Working paper.

H. Steckhalm et. al. (ed.) (1984). *Optimal Preisbildung under dynamischer Nachfrage*, 1984 (Berlin: Springer-Verlag).

Hanson, W. (2006). The dynamics of cost-plus pricing, *Managerial Decision Economics* **13**, 2, pp. 149–161.

Horsky, D. and Simon, L. (1983). Advertising and the diffusion of new products, *Marketing Science*, 2, pp. 1–17.

Jackson, C. and Pascual, R. (2008). Optimal maintenance service contract negotiation with aging equipment, *European Journal of Operational Research*, 189.

Jagpal, S. (2008). *Fusion for Profit. How Marketing and Finance Can Work Together to create Value* (Oxford University Press).

Jeuland, A. and Dolan, R. (1982). An aspect of new product planning: Dynamic pricing, *Management Science*, 18, pp. 1–21.

Johnson, W. C. and Bhatia, K. (1997). Technological substitution in mobile communication, *ournal of Business and Industrial Marketing,* **12**, 6, pp. 383–386.

Journal, E. (2008). IBM survey: 50 percent of consumers prefer mobile phones to PCS.

Keating, S. and Huff, K. (2005). Managing risk in the new supply chain [performance-based logistics], *Engineering Management Journal* **15**, 1, pp. 24–27.

Kim, N., Mahajan, V. and Srivastava, R. K. (1995). Determining the going market value of a business in an emerging information technology industry: The case of the cellular communications industry, *Technological Forecasting and Social Change*, 49, pp. 257–279.

Kim, S., Cohen, M. and Netessine, S. (2009). Reliability or inventory? Analysis of product support contracts in the defense industry, *Working Paper* .

Kim, S. H., Cohen, M. A. and Netessine, S. (2007). Performance contracting in after-sales service supply chains, *Management Science* **53**, 12, pp. 1843–1858.

Kim, Y. B., Seo, S. Y. and Lee, Y. T. (2000). A substitution and diffusion model with exogenous impact:forecasting of IMT-2000 subscribers in korea, *Vehicular Technology Conference, IEEE VTS 50th* **2**.

Kratz, L. (2008). Performance based life cycle product support – the new PBL, (W. Research (Ed.), North American Defense Logistis Conference, Crystal City Virginia: WBR Research), performance Based Life Cycle Product Support –The New PBL. In W. Research (Ed.), North American Defense Logistis Conference (2008). Crystal City Virginia: WBR Research.

Kreps, D. and Scheinkman, J. (1983). 'quality precommitment and Bertrand competition yield Cournot outcome, *Bell Journal of Economics* **14**, 2, pp. 326–337.

Kumar, V. and Petersen, J. A. (2008). Using a customer-level marketing strategy to enhance firm performance: A review of theoretical and empirical evidence, *Journal of the Academy of Marketing Science* **33**, 4.

Lenk, P. J. and Rao, A. G. (1990). New models from old: Forecasting product adoption by hierarchical Bayes procedures, *Marketing Science* **9**, 1.

Leo, P. (2006). Cell phone statistics that may surprise you, *Pittsburgh Post-Gazette* .

Levery, M. (2002). Making maintenance contracts perform, *Engineering Management Journal* **12**, 2, pp. 76 – 82.

Levesque, M. (2000). Effects of funding and its return on product quality in new ventures, *IEEE Transactions on Engineering Management* **47**, 1, pp. 98–105.

Libai, B., Muller, E. and Peres, R. (2009). The diffusion of services, *Journal of Marketing Research* **XLVI**, pp. 163–175.

Lilien, G. L. and Yoon, E. (1990). The timing of competitive market entry: An exploratory study of new industrial products, *Management Science* **36**, 5, p. 568.

Little, J. D. C. (1970). Models and managers: The concept of a decision calculus, *Management Science* **16**, pp. 466–485.

Mahajan, V. and Muller, E. (1979). Innovation diffusion and new product growth models in marketing, *Journal of Marketing* **43**, 4, pp. 55–68.

Mahajan, V., Muller, E. and Bass, F. M. (1990). New product diffusion models in marketing: A review and directions for research, *Journal of Marketing* **51**, pp. 1–36.

Mahajan, V., Muller, E. and Bass, F. M. (1995). Diffusion of new products: Empirical generalizations and managerial uses, *Marketing Science* **14**, 3.

Manning, K. C., Bearden, W. O. and Madden, T. (1995). Consumer innovativeness and the adoption process, *Journal of Consumer Psychology* **4**, 4, pp. 329–345.

Maoui, I., Ayhan, H. and Foley, R. D. (2009). Optimal static pricing for a service facility with holding costs, *European Journal of Operational Research* **197**, pp. 912–923.

Marn, M. V. and E. V. Roegner, C. C. Z. (2003). Pricing new products, *The McKinsey Quarterly, New York*, 3, pp. 40–49.

Marsland, N., Wilson, I., Abeyasekera, S. and Kleih, U. (2000). A methodological framework for combining quantitative and qualitative survey methods. An output from the DFID-funded natural resources systems programme (socio-economic methodologies component) project r7033 titled methodological framework integrating qualitative and quantitative approaches for socio-economic survey work, Tech. rep., The University of Reading.

Mesak, H. and Darrat, A. (2002). Optimal pricing of new subscriber services under interdependent adoption processes, *Journal of Service Research* **5**, 2, p. 140.

Meuter, M. L., Bitner, M. J., Ostrum, A. I. and Brown, S. W. (2005). Choosing among alternative service delivery modes: An investigation of customer trial of self-service technologies, *Journal of Marketing* **69**, pp. 61–83.

Movie Gallery Inc. (2008). Movie gallery inc. press release, URL `http://www.moviegallery.com/reorganization/default.html`.

Murthy, D. and Asgharizadeh, E. (1999). A stochastic model for service contracts, *International Journal of Reliability Quality and Safety* **5**, pp. 29–45.

Murthy, D. N. P., Rausand, M. and Virtanen, S. (2009). Investment in new product reliability, *Reliability Engineering and System Safety* **94**.

Murthy, D. N. P. and Yeung, V. (1995). Modelling and analysis of maintenance service contracts, *Mathematical and Computer Modeling* **22**, 10–12, pp. 219–225.

Murthy, D. P. and Blischke, W. R. (2006). *The Warranty Management and Product Manufacture* (Springer London).

Nader, F. H. and Jimenez, L. A. (2005). Substitution patterns, A Pitney Bowes Background paper for the project "Electronic Substitution for Mail: Models and Results, Myth and Reality", URL `http://www.postinsight.com`.

Nagle, T. T. (1987). *The strategy and tactics of pricing: A guide to profitable decision making* (Prentice Hall, New York, NY).

Neelamegham, R. and Chintagunta, P. (1999). A Bayesian model to forecast new product performance in domestic and international markets, *Marketing science* **18**, 2.

Nowicki, D., Kumar, U., Steudel, H. and Verma, D. (2008). Spares provisioning under performance-based logistics contract: Profit-centric approach, *Journal of the Operational Research Society* **59**, 3, pp. 342–352.

Olshansky, B. and Dossey, L. (2003). Retroactive prayer: a preposterous hypothesis, *BMJ* **327**, pp. 1465–1468.

Organization for Economic Co-operation and Development (2009). OECD Broadband portal statistics, URL http://www.oecd.org/document/54/0,3343.

Pontryagin, L. D. (1962). *The Mathematical Theory of Optimal Processes* (interscience, New York).

Preece, W. (1878). Quote, URL `www.quotes.net/quote/21123`.

Pride, W., Hughes, R. and Kapoor, J. (2008). *Business* (South-Western Cengago Learning).

Prins, R. (2008). *Modeling Consumer Adoption and Usage of Value-Added Mobile Services*, Ph.D. thesis, Erasmus University Rotterdam.

Program, N. C. R. (2009). Performance-based contracting for maintenance.

Randolph, N. (2007). Blockbuster to broaden entertainment reach, Content-AGenda Connecting Entertainment and Technology, URL `http://www.contentagenda.com`.

Rao, V. R. (1984). Pricing research in marketing: The state of the art, *Journal of Business* **57**, 1.

Reisinger, D. (2007). Say goodbye to blockbuster, URL `http://news.cnet.com/8301-13506_3-9809950-17.html?tag=mncol`.

Reisinger, D. (2008). Opinion: Can blockbuster be saved, URL `http://arstechnica.com/news`.

Robinson, B. and Lakhani, C. (1975). Dynamic price models for new product planning, *Management Science*, 21, pp. 1113–1122.

Rogers, E. M. (1962). *Diffusion of Innovations* (The Free Press, New York, NY).

Rogers, E. M. (1976). New product adoption and diffusion, *Journal of Consumer Research* **2**, pp. 290–301.

Rohlfs, J. (1974). Theory of interdependent demand for a communications service, *The Bell Journal of Economics and Management Science*, 5, pp. 163–175.

Roth, S., Woratschek, H. and Pastowski, S. (2006). Negotiating prices for customized services, *Journal of Service Research* **8**, 4, pp. 316–329.

Ruiz-Conde, E., Leeflang, P. S. H. and Wieringa, J. (2006). Marketing variables innmacro-level diffusion models, *JfB. State-Of-The-Art-Artikel*, 56, pp. 155–183.

Schubin, M. (2008). On the verge of... TVB Television Broadcast, URL www.televisionbroadcast.com/article/72032.

Seierstad, A. and Sydsaeter, K. (1977). Sufficient conditions in optimal control theory, *International Economic Review* **18**, 2.

Sherif, Y. and Smith, M. (1987). Optimal maintenance models for systems subject to failure -a review, *Naval Research Logistics Quarterly* **34**, pp. 47–74.

Sirius Satellite Radio (2010). News releases, .

Sols, A., Nowicki, D. and Verma, D. (2007). Defining the fundamental framework of an effective performance-based logistics contract, *Engineering Management Journal* **19**, 2, pp. 40–50.

Sols, A., Nowicki, D. and Verma, D. (2008). N-dimensional effectiveness metric-compensating reward scheme in performance-based logistics contracts, *Systems Engineering* **11**, 2, pp. 93–106.

Somerville, M. (1948). Quote, URL www.investorwords.com/tips/authors/1/mary-somerville.html.

Sound Partners Research (2008). URL http://www.soundpartners.ltd.uk/press_rel_wirelessVoIPforecasts.aspx.

Staiger, R. W. and Wolak, F. A. (1992). Collusive pricing with capacity constraints in the presence of demand uncertainty, *RAND Journal of Economics* **23**, 2, pp. 397–409.

Steenkamp, J. B., Hofstede, F. T. and Wedel, M. (1999). A cross-national investigation into the individual and national cultural antecedents of consumer, *Journal of Marketing* **63**, pp. 55–69.

Steenkamp, J. B. E. M. and Burgess, S. M. (2002). Optimum simulation level and exploratory consumer behavior in an emerging consumer market, *International Journal of Research in Marketing* **19**, pp. 131–150.

Stories, I. R. N. (2008). Netflix sails past $1 billion in e-commerce revenue in 2007, URL http://www.internetretailer.com/dailyNews.asp?id=25151.

Stremersch, S., Wuyts, S. and Frambach, R. T. (2001). The purchasing of full-service contracts: An exploratory study within the industrial maintenance market, *Industrial Marketing Management* **30**, 1, pp. 1–12.

TeleGeography, a research division of PriMetrica, Inc. (2008).

Tellis, G. J., Stremersch, S. and Yin, E. (2003). The international takeoff of new products: The role of economics, culture, and country innovativeness, *Marketing Science* **22**, 2, pp. 188–208.

Teng, J. and Thompson, G. (1983). Oligopoly models for optimal advertising when production costs obey a learning curve, *Management Science* **28**, 9, pp. 1087–1101.

Tseng, F. M. and Yu, C. Y. (2004). Partitioned fuzzyintegral multinomial logit model for taiwan's internet telephony market, *Omega* **33**, pp. 267–276.

US Census Bureau (2009). The 2009 Statistical Abstract.

Vitasek, K. and Geary, S. (2008). Performance-based logistics redefines department of defense procurement, *World Trade Magazine* **June**, pp. 62–65.

Warren, W. E., Abercombie, C. I. and Berl, R. (1989). Adoption of a service innovation: A case study with managerial implications, *The Journal of Service Marketing* **3**, 1, pp. 21–33.

XM Satellite Radio (2010). XM satellite radio releases.

Zhao, J., Wei, J., Sun, X. and Liu, J. (2008). Optimal pricing strategies with two substitutable products under decentralized decision, *Chinese Control and Decision Conference, 2008* .